John St. Loe Strachey

Dog stories

John St. Loe Strachey

Dog stories

ISBN/EAN: 9783744748711

Printed in Europe, USA, Canada, Australia, Japan

Cover: Foto ©berggeist007 / pixelio.de

More available books at **www.hansebooks.com**

PREFACE TO THE SECOND EDITION.

THE kindness with which "Dog Stories from the *Spectator*" has been received by the public, has made it necessary to issue a second edition. This affords me an opportunity of including some further examples of intelligence in dogs which have appeared in the *Spectator* since the formation of the original collection. Some of these new stories, it will I think be agreed, are quite as good as any of those previously published. Of very special interest are the anecdotes which I have placed together under the title of "Dogs and Human Speech." If we accept these stories as trustworthy, they seem to leave little doubt that many dogs are very far advanced in the understanding of human language. If, however, we reject this, the

1*

simple and straightforward explanation of
the evidence afforded by the letters, we must
choose one or other of the following solutions
of the problem.   These stories may of course
all be (1) hoaxes ; (2) instances of faulty
observation ; (3) coincidences, *i.e.*, the chance
linking of the utterance of certain words
with certain acts by dogs, though in reality
the words and acts had nothing to do with
each other ; (4) hallucination on the part of
the reporters—they were so convinced that
the dogs could understand, that they ima-
gined the dogs to do things which in reality
they never did.   Now, it must be observed
that each of these explanations would be,
primâ facie, perfectly tenable if there were
only one story of a dog understanding human
speech.   But it is difficult to see how any of
them will hold good for all the instances recorded, both in the new and in the old portion
of the present book.   It cannot seriously be
maintained that all the stories are hoaxes,
or that all the reporters were incapable
of relating what they saw, or, again, that
all the facts stated were due to mere co-

incidence or to hallucination. As far as I can see, the evidence shows that dogs do understand a good deal of human speech, though how they learn it remains a mystery. But in reality, and in practice, everyone admits this ability to understand human speech in the dog, though strenuous protests in the name of common-sense are made the moment anyone ventures to face the fact and put it boldly forward. Nobody doubts that a dog knows his own name and answers to it when called. But what is this but understanding human speech? If every dog can do this, why should not some dogs of exceptional quickness learn more than the one word, and so get a wider knowledge of human speech? Surely it is only the first word that matters. After that, knowing the whole dictionary is only a matter of degree. But though dogs seem able to understand our language, it is very curious to notice how utterly we fail to understand theirs. This point has been touched on by that able and eminent Judge of the Queen's Bench Division, Sir Henry

Hawkins. In a letter by him, printed in a little pamphlet which contains the biography of his fox-terrier Jack, Sir Henry Hawkins declares that his dog " understood all I said to him as though I spoke his own language, which not being accomplished enough to converse in I nevertheless thoroughly understood." Now, with all due deference to the learned Judge, I doubt very much whether he did understand Jack half as well as Jack understood him. If he did, he was certainly exceptionally fortunate. Men who own dogs no doubt get to know roughly what a bark at this or that time is likely to mean, but they seldom go further than this. Who ever heard two dogs barking together and understood that Jack was saying to Grip, " I mean to run away to-morrow ? " Yet as much as this ought to be understood by the man who would understand dog talk, as well as the probationary fox-terrier in the story headed " The Dog that heard he did not give satisfaction "—a story which is to be found among the new anecdotes at the end of the present volume.

Depend upon it, even Sir Henry Hawkins has not yet reached that pitch of knowledge, and though he doubtless understood Jack better than anyone else, I feel sure that Jack, as I have said, took in much more of his conversation than he did of Jack's. But it is far easier to point out our ignorance of dog language, than to suggest a means for surmounting that ignorance. Dogs are not like apes or rooks, of a conversational turn, and though they clearly communicate a great deal to each other, it is difficult to see how their language is to be studied. I doubt if the American professor's plan would be of much use, or if phonography in the kennels would give us any facts to work on.

Among the new stories in the present edition will be found a further series illustrating the power of dogs to feel the emotions of grief and of devotion for each other. There are also some interesting letters making up a fairly complete biography of Bob, the Australian Railway dog. Under the heading "More Miscellaneous," will be found some exceedingly strange and amusing stories. The story of the Religious Dog and the Pagan Cat is quite excellent as a piece of humour, while the Praying Dog suggests a strange picture of sanctimonious canine priggishness. Crib's biography is also not a little curious. I have only to add my reason for not distributing the new stories as far as possible under their appropriate headings in the body of the book. This would, I admit, have been the logical and natural course. There were, however, two objections. First, there exist, I am given to understand, typographical difficulties which, if not insurmountable, are serious. Next, a reader of the first edition who might come across the second and wish to see what was fresh, would be much puzzled to get at the new stuff. It would therefore probably be his wish that the added stories should be grouped together at the end. But a reader of the first edition is clearly an old friend, and must have his wishes consulted, especially when doing so does not interfere with the enjoyment of him who reads for the first time in the second issue of these stories.

INTRODUCTION.

I.

THE following Dog Stories are taken from
the pages of the *Spectator*, with the permission of the editors and proprietors.
It was suggested to me by Mr. Fishei
Unwin that the many strange and pleasant
stories of dogs which from time to time
are sent to the *Spectator* by its correspondents would, if put together, form a
volume of no little entertainment for all
who love dogs, or are interested in stories
of animal intelligence. Up till now the
Spectator dog stories, after the week of
their publication, have practically been in-
accessible to the general reader ; for he is a
bold man who will attack a bound volume
of a newspaper in search of amusement.
Though I at once agreed that the suggested

book would be a very readable one, and
likely to please dog-lovers all the world
over, I did not, till the selection was nearly
made, realise how much the stories gain by
being grouped together. A single story
of a clever dog may amuse, but it is liable
to be put aside as an accident, a coincidence,
a purely exceptional circumstance which
proves nothing. If, however, instead of
a single story we have half a dozen illustrating the same form of intelligence, the
value of the evidence is enormously increased, and a collection of dog stories
may become of very great value in deter-
mining such questions as the power of dogs
to act on reason as well as on instinct,
or their ability to understand human language. The solution of these problems is, I
cannot help thinking, materially advanced
by the stories in the present book. Take,
again, the group of stories which I have
labelled Purchasing Dogs. One sample of
this kind might, as I have noted above, be
put off as a case of imperfect observation,
or as a curious coincidence ; but when we

get a whole group of stories it becomes very difficult to doubt that dogs may learn the first principles of the science of exchange. The Italian dog (page 59) which did the narrator a service by fetching him cigars, demanded payment in the shape of a penny, and then used that penny by exchanging it for a loaf, was far advanced in the practice of Political Economy. He not only understood and acted on an implied contract, but realised the great fact at the back of the currency. " What are guineas," said Horne Tooke, " but tickets for sheep and oxen !" The Italian dog did not, like a savage, say, " What is the use of copper to me, I cannot eat it?" Instead, he perceived that the piece of copper was a ticket for bread. It should be noted too that this dog, the dog called Hardy (page 57) and others, were able to distinguish between the pieces of copper given them. Again, the Glasgow story (page 53) shows that a dog can learn to realise that a halfpenny will buy not merely one thing but several things—in fact, that the great advantage of exchange by currency over barter is that it gives you a choice. While on the subject of purchasing dogs, it is curious to reflect how very little is wanted to convert the dog that is able to purchase into a free agent. If a dog can exchange his faculty for cigar carrying or his tricks against half-pence, why should he not exchange useful services, such as guarding a house or herding sheep, and so become self-supporting? Imagine a collie paid by the day, and, when his work was over, receiving twopence and going off to buy his supper. But the vista opened is too far-reaching. One sees down it dogs paid by the hour and by the piece, and then dogs asking for better pay and shorter hours, and, finally, dogs on strike, and dog " black-legs," or " free dogs."

II.

A word should be said as to the authenticity of the stories in the present volume. It is a matter of common form for the evening newspapers to talk of the *Spectator* dog stories as hoaxes, and to refer in their

playful, way to "another *Spectator* dog."
It might not then unnaturally have been
supposed that a person undertaking to edit
and reprint these stories would have found
a considerable number that showed signs
of being hoaxes. I may confess, indeed,
that I set out with the notion of forming
a sort of Appendix to the present work,
which should be headed "Ben Trovato,"
in which should be inserted stories which
were too curious and amusing to be left
out altogether, but which, on the other
hand, were what the Americans call a little
"too tall" to be accepted as genuine. The
result of my plan was unexpected. Though
I found many stories in which the inferences
seemed strained or mistaken, and others
which contained indications of exaggeration,
I could find but two stories which could
reasonably be declared as only suitable for
a "Ben Trovato." I therefore suppressed
my heading. The truth is that the animal
stories are much more carefully sifted at
the *Spectator* office than our witty critics
and contemporaries will admit. No stories

are ever published unless the names and
addresses of the writers are supplied, and all
stories are rejected which have anything
clearly suspicious about them. What the
editors of the *Spectator* do not do is to reject
a dog-story because it states that a dog has
been observed to do something which has
never been reported as having been done
by a dog before, or at any rate, something
which is not universally admitted to be doable
by a dog. Apparently this willingness to
print stories which enlarge our notions of
animal intelligence is regarded in certain
quarters as a sign that the *Spectator* will
swallow anything, and that its stories must
be apocryphal. I cannot, however, help
thinking that all who care for the advance-
ment of knowledge in regard to animals
should be grateful to the editors of the
Spectator for not adopting the plan of ex-
cluding all dog stories that do not correspond
with an abstract ideal of canine intelligence.
Had they acted on the principle of putting
every anecdote that seemed *primâ facie* un-
likely into the waste-paper basket, they would

certainly have missed a great many stories of real value. In truth, there is nothing so credulous as universal incredulity. An attitude of general incredulity means a blind belief in the existing state of opinion. If we believe that animals have no reasoning power, and refuse to examine evidence that is brought to show the contrary, we are adopting, the attitude of those who disbelieve that the earth goes round the sun because they seem daily to see a proof of an exactly opposite proposition. If people are to refuse to believe anything of a dog that does not sound likely on the face of it, we shall never get at the truth about animal intelligence. What is wanted is the careful preservation and collection of instances of exceptional intelligence.

III.

Before I conclude this Introduction, I should like to address a word of apology to the correspondents of the *Spectator* whose letters form the present volume. Though the copyright of the letters belongs to the editors and proprietors of the *Spectator* I should have liked to ask the leave of the various writers before republishing their letters. Physical difficulties have, however rendered this impossible. In the case of nearly half the letters the names and addresses have not been preserved. In many instances, again, only the names remain. Lastly, a large number of the letters are ten or twelve, or even twenty years old, and the writers may therefore be dead or out of England. Under these circumstances I have not made any effort to enter into communication with the writers before including their letters in this book. That their permission would have been given, had it been asked, I do not doubt. The original communication of the letters to the *Spectator* is proof that the writers wished a public use to be made of the anecdotes they relate. As long, then, as the letters are not altered or edited, but produced verbatim, I may, I think, feel assured that I am doing nothing which is even remotely discourteous to the writers.

SYLLOGISTIC DOGS.

A DOG ON LONG SERMONS.

[*Aug.* 4, 1888.]

DURING a recent journey in Canada, I met with a striking instance of reason in a dog. I was staying at the Mohawk Indian Institution, Brantford, Ontario. The Rev. R. Ashton, superintendent of the school, is also incumbent of the neighbouring Mohawk Church (the oldest Protestant church in Canada). Mr. Ashton is very fond of animals, and has many pets. One of these, a black-and-tan terrier, always accompanies the ninety Indian children to church on Sunday morning. He goes to the altar-rails, and lies down facing the congregation. When they rise to sing, he rises; and when they sit, he lies down. One day, shortly before my visit, a stranger-clergyman was preaching, and the sermon was longer than usual. The dog grew tired and restless, and at last a thought occurred to him, upon which he at once acted. He had observed that one of the elder Indian boys was accustomed to hand round a plate for alms, after which the service at once concluded. He evidently thought that if he could persuade this boy to take up the collection, the sermon must naturally end. He ran down to the back seat occupied by the boy, seated himself in the aisle, and gazed steadfastly in the boy's face. Finding that no notice was taken, he sat up and "begged" persistently for some time, to Mr. Ashton's great amusement. Finally, as this also failed, the dog put his nose under the lad's knee, and tried with all his strength to force him out of his place, continuing this at intervals till the sermon was concluded.

Did not this prove a distinct power of consecutive reasoning?

A. H. A.

A COMMERCIAL TREATY
BETWEEN A DOG AND A HEN.

[*July* 7, 1888.]

YOUR dog-loving readers may be interested in the following instance of animal sagacity. Bob is a fine two-year-old mastiff, with head and face of massive strength, heightened by great mildness of expression. One day he was seen carrying a hen, very gently, in his mouth, to the kennel. Placing her in one corner, he stood sentry while she laid an egg, which he at once devoured. From that day the two have been fast friends, the hen refusing to lay anywhere but in " Bob's " kennel, and getting her reward in the dainty morsels from his platter. There must have been a bit of canine reasoning here. " Bob " must have found eggs to his liking, that they were laid by hens, and that he could best secure a supply by having a hen to himself.

THOMAS HAMER.

A DOG NURSE.

[*Feb.* 20, 1875.]

A PATIENT recently consulted me who was blind and subject to fits. I pointed out to her friends the danger to which she was exposed in case a fit came on when she was in the vicinity of a fire, and they informed me that she incurred little or no risk, because a favourite dog ran at once and fetched assistance the moment a fit came on. This intelligent animal would rush into the next house barking eagerly, would seize the dress of the woman who lived there, and drag her to the assistance of his mistress. If one did not go, he would seize another, and exhibited the most lively symptoms of distress until his object was accomplished.

CHARLES BELL TAYLOR, M.D., F.R.C.S.

INSTINCT, OR REASON?

[*Sept.* 1, 1888.]

THE following incident in dog-life may perhaps find a place in the *Spectator*. I quote
from a letter received a few days ago from
my nephew, " T. G. T.," resident in South
Africa :—" Johannesburg, Traansvaal.—My
dog Cherry has had three great pups, and
I had to leave her behind at the Grange.
When I was going away, Cherry and the
pups were located in some stables. She
came out and watched the tent-truck and
my things packed up. Presently I went
away, and when I came back I found
Cherry had carried all the pups on to the
top of my luggage, and evidently had not
the least intention of staying behind."

T. W. T.

HOSPITAL DOGS.

[*June* 26, 1875.]

DR. WALTER F. ATLEE writes to the editor of the *Philadelphia Medical Times :—*

"In a letter recently received from Lancaster, where my father resides, it is said :— 'A queer thing occurred just now. Father was in the office, and heard a dog yelping outside the door; he paid no attention until a second and louder yelp was heard, when he opened it, and found a little brown dog standing on the step upon three legs. He brought him in, and on examining the fourth leg, found a pin sticking in it. He drew out the pin, and the dog ran away again.' The office of my father, Dr Atlee, is not directly on the street, but stands back, having in front of it some six feet of stone wall with a gate. I will add, that it has not been possible to discover anything more about this dog.

"This story reminds me of something similar that occurred to me while studying medicine in this same office nearly thirty

years ago. A man, named Cosgrove, the
keeper of a low tavern near the railroad
station, had his arm broken, and came many
times to the office to have the dressings
arranged. He was always accompanied by
a large, most ferocious-looking bull-dog, that
watched me most attentively, and most un-
pleasantly to me, while bandaging his
master's arm. A few weeks after Cosgrove's
case was discharged, I heard a noise at the
office door, as if some animal was pawing it,
and on opening it, saw there this huge bull-
dog, accompanied by another dog that held
up one of its front legs, evidently broken.
They entered the office. I cut several
pieces of wood, and fastened them firmly
to the leg with adhesive plaster, after
straightening the limb. They left immediately. The dog that came with Cosgrove's
dog I never saw before nor since."

Do not these stories adequately show that
the dogs reasoned and drew new inferences
from a new experience ?

B.

[*April* 6, 1889.]

KNOWING your interest in dogs, I venture to send you the following story. A week or two ago, the porter of the Bristol Royal Infirmary was disturbed one morning about 6.30 by the howling of a dog outside the building. Finding that it continued, he went out and tried to drive it away; but it returned and continued to howl so piteously, that he was obliged to go out to it again. This time he observed that one of its paws was injured. He therefore brought it in and sent for two nurses, who at once dressed the paw, and were rewarded by every canine sign of gratitude, including much licking of their hands. The patient was "retained" for two days, during which time he received every attention from those inside the house, and from the neighbours outside, who quickly heard of the case. As no one appeared to claim the dog, he was sent to the Home for Lost Dogs in the city, where so interesting an animal was, of course, not long in finding a purchaser. The dog was one of those called "lurchers."

I have myself called on the porter of the infirmary for confirmation of the story, and am assured by him of its truth. How did an apparently friendless dog know where to go for surgical aid ? The case differs from that of the dog which took its friend for treatment to King's College Hospital in London, for I understand that the King's College dog had previously been taken to the hospital for treatment itself ; but in this case there is no such clue.

HELEN M. STURGE.

FEATURES IN THE CHARACTER OF A DOG.

[*June* 10, 1876.]

FOR some time past I have noticed in your journal letters and articles referring to the wonderful powers of dogs. As I was myself much struck by many features in the character of a dog which I knew, illustrating, as I think, not only affection, but reasoning faculties, I shall acquaint you with a few of these, believing that they may be *interesting*, at least to all admirers of that noble animal.

The dog of which I speak was a terrier. It showed its affection in the most marked manner in several ways. Every morning, as soon as it got out of the kitchen, it came to its master's door, and if not admitted and caressed about the usual hour, gave evident signs of impatience. It would lie quiet till it thought the time had arrived, but never longer. Afterwards it went to the breakfast-room, and occupied its master's chair till he arrived. On one occasion a visitor was in the house, who, coming first into the room, ordered the dog to come off the best chair. To this it paid no attention, and when threatened with expulsion, at once prepared for defence.   But as soon as its master appeared it resigned its place voluntarily, and quietly stretched itself on the rug at his feet.

At another time it was left for three weeks during its master's absence from home.   It saw him leave in a steamer, and every day until his return it repaired to the quay upon the arrival of the same boat, expecting him to come again in the one by which he had gone.   It distinguished between a number of boats, always selecting the right one and the right hour.

One evening it accompanied its master when he went to gather mussels for bait. As the tide was far in, few mussels remained uncovered ; and after collecting all within reach, more were required.   A large bunch lay a few feet from the water's edge, but beyond reach ; yet as the dog was not one of those who take the water to fetch, its master had no expectation that it would prove useful on the present occasion. Seeing him looking at the mussels, however, it first took a good look at those in the basket, and then, without being directed at all, went into the water. Selecting the right bunch from amongst the stones and wreck with which it was surrounded, it brought it to land, and laid it at its master's feet. This, I think, is a proof of *reason*, rather than of instinct. The dog had never been trained to go into the sea, and would not probably have brought out the mussels had it not seen that they were wanted.

It showed wonderful instinct, however, just before the death of one of its pups, and before its own death. Its pup had not been thriving, and the mother gave unmistakable proof that she foresaw its death. She dug a grave for it and put it in. Nor, when it was removed, would she let it lie beside her, but immediately dug another grave, where she was less likely to be disturbed. Upon the day of her own death, also, she used what strength she had to dig her grave, in which she lay, preferring to die in it, than in what would seem to most a place of greater comfort.[1]

These may not be singular incidents, but they are still remarkable and worthy of notice. They serve to show us the wonderful nature of man's faithful friend, the dog, and how he has many traits of character fitted to make him the worthy receiver of kindness and respect.

T.

[1] It is difficult to accept T.'s explanation of the dog's object in digging. Possibly its aim was to obtain warmth or shelter.

BULLY'S SHORT CUT.

[*Aug.* 29, 1874.]

I SEE that you welcome all notes of interest upon our fellow-beings, the dogs. Here is one that seems to prove they have a sense of time and of distance as measured by time.

I was walking with my bull-terrier, Bully (seven years old last Christmas), during a hot afternoon this month homewards along the Bund (Shanghai), and I suddenly missed him. I turned back for twenty or thirty yards, and, not finding him, I gave up the search, saying, " He knows the way home well enough." Presently I saw him on my right, dripping with water, cantering on at a round pace, without looking about him, homewards. I watched him, curious to see whether he would go straight home. No. He kept on till he reached the distance of about 150 yards, and looked ahead, *not* smelling the ground. He then deliberately walked back, catching sight of me in about twenty yards after his turning back, and wagged his tale recognisingly. He had evidently been to cool himself in the river (thirty yards to the right, it being low tide), and, thinking I would go on at the ordinary pace without him, he, after his bath, struck directly at a long diagonal for the point I would have reached if I had not turned back to look for him. He did not seem to have the slightest misgiving as to his sense of the distance I *ought* to have walked during the time of his bath. His turning was done seemingly with a calm assurance of certainty. I may add that there were twenty to thirty foot-passengers scattered over the portion of road in question at the time, whose footsteps might have effaced my scent on the *watered* granite macadamised roadway, even supposing the dog to have tried his sense of smell, *which he did not*, as far as I could see, and I noticed him carefully.

W. G. S.

CANINE INTELLIGENCE.

[*July* 24, 1886.]

You often give us pleasant anecdotes of our four-footed friends. You may think the following worthy of record. I have a little dog, a not particularly well-bred fox-terrier. He is much attached to me, and shows by his obedience, and sometimes *in* his disobedience, that he understands a good deal. Yesterday I was away all day, and he, I am told, was very uneasy, and searched everywhere for me. Every day at 5 p.m. I go to church. Toby seems to know this is not an ordinary walk, and never offers to come with me. But yesterday, when the bell began, he started off and took up his position by the vestry door. I believe he reasoned with himself, " There goes the bell; now I shall catch the Vicar."

WILLIAM QUENNELL.

THE DOG AND THE FERRY.

[*April* 4, 1885.]

READING from time to time many pleasant anecdotes in the columns of the *Spectator*— which, by the way, I receive as regularly, and read as eagerly, as when resident in England many years ago—relative to the sagacity of dogs, I send the following, thinking it possible you may deem it worthy of insertion.

Some three years ago I was "having a spell" in Brisbane, after a lengthened sojourn on a sheep station in the interior of Queensland. During my stay in the city I had the good fortune to gain the friendship of a gentleman who owned a magnificent collie. My friend, his dog Sweep, and myself, were frequently together, engaged either in yachting among the islands of Moreton Bay, or 'possum hunting under the towering *eucalypti* which fringe the banks of the river Brisbane. Naturally "Sweep" (who was a most lovable animal) and myself soon began to entertain a warm friendship for one

3

another, which friendship gave rise to the
anecdote I am about to relate. Returning
to my hotel about midnight from the house
of a friend, I was not a little startled at finding my hand suddenly seized from behind by
a dog, which, however, I at once recognised
as my handsome acquaintance, Sweep. I
patted him, at the same time endeavouring
to withdraw the hand which he held firmly,
but gently, between his teeth. It was of no
use, as, in spite of all my endearments, he
insisted on retaining his hold, wriggling along
by my side, and vigorously wagging his tail,
as though he would say, " Don't be afraid;
it's all right." We soon reached a point in
the main street down which we were walking, where a side avenue branched off
towards the river. My way lay right ahead.
Sweep, however, insisted on my taking
the road which lay at a right-angle to my
course. I felt some annoyance at his persistence, as I was both tired and sleepy;
but, having no choice in the matter, I
followed his lead. Having walked some two
or three hundred yards down *his* street, he

released his hold, dancing round me, then running on for a few yards and looking back to see if I were following. Becoming interested, I determined to see what he was after, so, without further resistance, I followed submissively. At last, having reached the river, which at this place was about four hundred yards wide, he, with many joyous barks, ran down the ferry steps, and jumped into the empty boat of the ferryman. At last I was able to guess at his motive for forcing me to follow him. His master, who lived across the river, had accidentally lost sight of his dog returning from his office in the city; and Sweep appeared to understand perfectly that unless the boatman received his fare he, Sweep, would not be carried over, my friend frequently sending the dog over by himself when wishing to attend concerts, &c., invariably paying the fare as of an ordinary passenger. The ferryman, who at once recognised my canine friend, laughed heartily when I told him how I had been served, took my penny, and set off at once for " Kangaroo Point," Sweep gaily barking "good-night" until he reached the opposite bank. I heard subsequently that he used to swim the river when left behind; but having had two narrow escapes from sharks, his nerves had become somewhat shaken so far as water was concerned.

J. Wм. Creighton.

THE REASON OF DOGS.

[*Nov.* 13, 1875.]

HAVING often read, with great pleasure, the anecdotes about dogs which from time to time appear in the *Spectator*, I venture to send you one which has come under my own observation, and which, it seems to me, shows an effort of reasoning implying two distinct ideas—one the consequence of the other—more interesting than many of those clever performances of educated dogs which may or may not be merely mechanical actions.

The dog who performed the following trick was then a great, half-grown, awkward puppy, whose education, up to that time, had been much neglected. It has been better attended to since, and now, although sportsmen probably consider such an animal sadly thrown away upon a lady, he is a very pleasant friend and companion. My two dogs, Guy and Denis, form as capital a pair, for contrast's sake, as one need wish to see. They are both handsome dogs of their kind

—Guy, a fine black retriever, with no white hair upon him, and, I believe, in the eyes of sportsmen, as well as those of his mistress, a very desirable possession, good-tempered, clever, and affectionate ; Denis, as naughty and spoilt a little fellow as ever existed, and a great pet, also black, except for his yellow paws and chest, but covered with long, loose locks, instead of Guy's small, crisp curls.

Denis is exceedingly comic, and a constant source of amusement. He is very faithful to his mistress, whose bedside during illness he has refused to leave, even for food; but it must be confessed that he is not amiably disposed towards most people, and is a perfect tyrant over the other animals. Some account of the two dogs' character is necessary, to explain the little scene which took place between them one evening about a year ago. Guy, it must be premised, is at least twelve months younger than Denis, consequently, when the former first arrived—a miserable and very ugly little puppy, a few weeks old, more like a small black jug than any known animal of the canine species,

having had the mange, and lost all his hair—
Denis undertook his education, and ruled
him so severely that his influence lasted a
long while; indeed, even after Guy had
grown so big that Denis almost needed to
stand upon his hind legs in order to snap
at him, the great dog would crouch meekly
at a growling remonstrance from the little
master, and never dared to invade his rights
—to approach his plate of food, or to drink
before him. Now a days Guy has discovered his own power, and although too
good-natured an animal ever to ill-treat the
little dog, no longer allows any liberties, but
at the same time, when the scene which I
am about to describe took place, he was still
under the impression that Denis's wrath was
a terrible and dangerous matter.

And now for my story, which, it seems to
me, shows as much real reasoning power in
an untrained animal as any anecdote that I
ever read. One evening I took my two
dogs to the kitchen, to give them the rare
treat of a bone apiece. (Dogs were certainly
never intended to make Natal their home,

for, in order to keep them alive at all, they should never be given anything that they like, especially meat, and even then the most careful management often fails in preserving them from disease and death.)   One of my sisters was with me, and together we watched the dogs over their supper.   Guy, with his great mouth, and ravenous, growing appetite, made short work with his, every vestige of which had vanished ; while little Denis was still contentedly sucking away at his small share, not very hungry, and taking his pleasures sedately, like a gentleman, as he is. And then Guy began to watch the other with an envious eye, evidently casting about in his mind how he might gain possession of that bone.   He was even then, though not full grown, so big and strong that he could have taken it by force with the greatest ease ; but such an idea did not cross his mind ; he decided to employ stratagem to win the prize. I must mention here, that amongst other naughty practices of my dogs, is that of rushing out of the house and barking violently upon the slightest sound without.

This is Denis's fault, which Guy, in spite of
all my lessons, has contracted from him.
With the evident intention of sending Denis
out, Guy suddenly started up, and began to
bark *towards* the door in an excited manner,
but not running out himself, as he certainly
would have done, had he really heard anything. Down went Denis's bone, and out
rushed he, barking at the top of his voice.
Did Guy follow him? Oh, dear no! he had
no such intentions; he sneaked up to Denis's
bone immediately, picked it up, and ran to
the other end of the room. But when he
had got it, he did not know what to do with
it; there was no hiding-place for him there,
and he dare neither await Denis's return
openly, nor risk meeting him at the door.
My sister and I were, by this time, both
sitting on a bench against the wall, watching
the scene between the dogs, and Guy, after
running once round the room, with the bone
in his mouth, came and crept in beneath my
seat, where he was hidden by my dress, and
where he lay, not eating the bone, and in
perfect silence. Presently Master Denis

trotted back, quite unconscious, and shaking
the curls out of his eyes, as much as to say,
"My dear fellow! what a fuss you've made;
there's nothing there." He looked about for
his bone for a few minutes, but soon gave up
the search, and began to amuse himself with
other things. After a while, I, forgetting the
culprit beneath my seat, rose, and crossed the
room, leaving him exposed. Guy was in a
great fright; he jumped up, and running to
my sister, who was still seated, he stood up
with his forepaws upon her lap, and the
bone still untouched in his mouth, as though
begging her protection. Denis, however, did
not observe him, and after a few minutes,
Guy's courage returned, and finally he
ventured to lie down, with the bone between
his paws, and began to gnaw it, keeping one
eye fixed on Denis the while. This, however, was going a step too far. Denis was
attracted by the sound, and recognised his
own bone the moment that he looked round.
He marched up to Guy (who immediately
stopped eating) and stood before him.
Denis growled, and Guy slowly removed

one great paw from his prize. Denis advanced a step, with another growl; Guy removed the other paw, and slunk back a little, whereupon Master Denis calmly walked up, took possession of his bone, and went off with it.

I am bound, however, to remark that after another half-hour's contented amusement over it, he resigned the remainder, which was too hard for his small mouth, to Guy, who finished the last morsel with great satisfaction. Now that he is full grown, Guy still gives up to Denis in many little ways, but it is evidently through generosity only, for he has proved himself perfectly capable of taking his own part. But he is very gentle with his little playmate, except at night, when he lies across my door-way—entirely of his own accord—and will allow no one and nothing to enter without my command.

FRANCES E. COLENSO.

A CANINE SIGHT-SEER.

[*May* 20, 1876.]

As a subscriber to your journal, I have observed from time to time discussion on the " reasoning power of dogs." I will tell you what I observed to-day. In consequence of the Levée there was a great crowd in Pall Mall. I was invited by a friend to accompany him in his carriage from St. James's Palace down Pall Mall, when lo and behold, his dog, which usually runs with the carriage, insisting on getting in also. Nothing could induce him to get out, and whilst passing along Pall Mall he amused himself looking out of window at the police, soldiers, and crowd collected. When through, he was glad enough to get out again, and readily followed through the most frequented streets. Now, I have no doubt as to that dog's " reasoning power," respecting his ability to follow his carriage safely through the dense crowd collected around St. James's Palace and Marlborough House.

H.

THINKING OUT A PLAN.

[March 3, 1888.]

ARE animals able to think over and carry out a plan? The following anecdotes will answer the question. When in India, I had a small rough terrier who, when given a bone, was sent to eat it on the gravel drive under an open porch in front of the bungalow. On several occasions two crows had made an attempt to snatch the dainty morsel, but their plans were easily defeated by Topsy's growls and snapping teeth. Away flew the crows to the branch of a tree near by. After a few moments of evident discussion, they pro ceeded to carry out the plan of attack. One crow flew down to the ground and gave a peck at the end of the dog's tail. Topsy at once turned to resent this attack in the rear, whilst the other crow flew down and bore the bone away in triumph.

The same dog had a favourite resting-place in an easy-chair, and was very often deprived of it by a dog which came as visitor to the house. Topsy did not approve of this, and her attempts to regain her seat were met with growls and bites. This justified an act of eviction, and the busy little brain decided on a plan. The next day, as usual, the intruder established himself in the chair, which was close to the open door. Topsy looked on for a moment, and then flew savagely out of doors, barking at a supposed enemy. Out ran the other dog to see what was up, and back came Topsy to take possession of her coveted seat. The other dog came slowly back, and curled himself up in a far-off corner. The above I was an eye-witness to, and therefore can vouch for the truth of what I relate.

K. P.

A PARCEL-CARRYING DOG.

[Feb. 9, 1895.]

In illustration of the anecdotal letters about dogs and their habits, in the *Spectator* of February 2nd, and Mr. Lang's paper in this month's *Nineteenth Century*, I send you the following story of a dog which I had in 1851 and for three years afterwards. He was a handsome Newfoundland dog, and one of the most intelligent animals with which it was ever my good luck to meet. I was living in a village about three miles from Dover, where I did all my shopping and marketing, being generally my own "carrier." Sometimes Nep would carry home a small parcel for me, and always most carefully. On one occasion Nep was with me when I chose a spade, and asked the ironmonger to send it by the village carrier. The spade was put by, labelled and duly addressed. I went on to have a bathe, my dog going with me, but on finishing my toilet in the machine, and calling and whistling for Nep, he was nowhere to be seen. He was not to be found at the stable where I had left my horse, but on calling at the ironmonger's shop I found he had been there and had carried off the spade which I had bought, balancing it carefully in his mouth. When I reached home, there Nep was, lying near his kennel in the stable-yard looking very fagged, but wearing a countenance of the fullest self-satisfaction, and evidently wishing me to think he had fulfilled his "dog-duty." My friend Mr. Wood, who was a thorough lover and admirer of dogs, was delighted to hear of his intelligent performance.

"CANOPHILIST."

P.S.—I may add Nep always guarded me when bathing, and always went into the water with me, too, often uttering a peculiar kind of "howl."

THE REASONING POWERS OF DOGS.

PURCHASING DOGS.

[*May* 26, 1877.]

SOME time ago I sent you my recollections of
a dog who knew a halfpenny from a penny,
and who could count up as far as two (see page
56). I have been able to obtain authentic
information of a dog whose mental powers
were still more advanced, and who, in his
day, besides being celebrated for his abilities,
was of substantial benefit to a charitable
institution in his town. The dog I refer to
was a little white fox-terrier, Prin by name,
who lived at the Lion Hotel, at Kidderminster, for three or four years ; but now,
alas ! he is dead, and nothing remains of him
but his head in a glass case.

I had heard of this dog some months ago,
but on Saturday last, having to make a visit
to Kidderminster, I went to see him. The
facts I give about him are based on the

statements of Mr. Lloyd, his master, and
they are fully substantiated by the evidence
of many others. I have before me a state-
ment of the proceeds of " Dog Prin's box,
Lion Hotel ; subscriptions to the Infirmary."
The contributions began in September, 1874,
and ended on April 25th, 1876, and during
that period the sum of £13 14s. 6d. was
contributed through Prin's instrumentality.

He began by displaying a fancy for playing with coins, not unusual amongst terriers,
and he advanced to a discovery that he
could exchange the coins for biscuits. He
learnt that for a halfpenny he could get two
biscuits, and for a penny, three ; and, having
become able to distinguish between the two
coins, it was found impossible to cheat him.
If he had contributed a penny, he would not
leave the bar till he had had his third biscuit;
and if there was nobody to attend to his
wants, he kept the coin in his mouth till he
could be served. Indeed, it was this persistence which ultimately caused poor Prin's
death, for there is every reason to fear that
he fell a victim to copper-poisoning.

By a little training he was taught to place the coins, after he had got the biscuits, upon the top of a small box fixed on the wall, and they were dropped for him through a slot. He never objected to part with them in this way, and having received the *quid pro quo*, he gave complete evidence of his appreciation of the honourable understanding which is so absolutely necessary for all commercial transactions.

An authenticated case like this is of extreme value, for just as the elementary stages of any science or discovery are the most difficult and the slowest in accomplishment, so are the primary stages of all mental processes. To find the preliminary steps of the evolution of mathematics and commerce in a dog is therefore a very important observation, and everything bearing on these early phases of intellect should be carefully recorded.   LAWSON TAIT.

[Feb. 10, 1877.]

THE *Spectator* is always so kind to animals that I venture to send you the following story of a dog's sagacity, which may be depended upon as absolutely true :—

During the meeting of the British Association at Glasgow, a friend of mine had occasion to go one day from that place to Greenock on business. Hearing, on his arrival, that the person he wished to see was out, but expected shortly to return home, he determined to take a stroll about the town, to which he was a stranger. In the course of his walk he turned into a baker's shop and bought a bun. As he stood at the door of the shop eating his bun, a large dog came up to him and begged for a share, which he got, and seemed to enjoy, coming back for piece after piece. "Does the dog belong to you?" my friend asked of the shop-woman. "No," she answered, "but he spends most of his time here, and begs halfpennies from the people who pass.' "Halfpennies! What good can they be to him?" "Oh, he knows very well what to do with them; he comes into the shop and buys cakes."

This seemed rather a remarkable instance

of cleverness even for the cleverest of animals, so, by way of testing its reality, my friend went out of the shop into the street, where he was immediately accosted by the dog, who begged for *something* with all the eloquence of which a dog is capable. He offered him a halfpenny, and was rather surprised to see him accept it readily, and walk, with the air of a regular customer, into the shop, where he put his forepaws on the counter, and held out the halfpenny towards the attendant. The young woman produced a bun, but that did not suit the dog, and he held his money fast. " Ah," she said, " I know what he wants," and took down from a shelf a plate of shortbread, This was right ; the dog paid his halfpenny, took his shortbread, and ate it with decorous satisfaction. When he had quite finished he left the shop, and my friend, much amused, followed him, and when he again begged found another halfpenny for him, and saw the whole process gone through a second time.

This dog clearly had learned by some means the use of money, and not merely that it would buy something to eat, but that it would buy several things, among which he could exercise a right of choice. What is perhaps most remarkable is that his proceedings were entirely independent, and for his own benefit, not that of any teacher or master. A. L. W.

[*Feb.* 17, 1877.]

WHEN a student at Edinburgh, I enjoyed the friendship of a brown retriever, who belonged to a fishmonger in Lothion Street, and who was certainly the cleverest dog I have ever met with. He was a cleverer dog than the one described by "A. L. W." because he knew the relative value of certain coins. In the morning he was generally to be seen seated on the step of the fishmonger's shop-door, waiting for some of his many friends to give him a copper. When he had got one, he trotted away to a baker's shop a few doors off, and dropped the coin on the counter. If I remember rightly (it is twelve or fifteen years ago), his weakness was "soda scones." If he dropped a halfpenny on the counter he was contented with one scone, but if he had given a penny he expected two, and would wait for the second, after he had eaten the first, until he got it. That he knew exactly when he was entitled to one scone only, and when he ought to get two, is certain, for I tried him often.

LAWSON TAIT.

[*Feb.* 17, 1877.]

IN the *Spectator* of the 10th inst. a correspondent describes the purchase of cakes by a clever dog at Greenock. I should like to be allowed to help preserve the memory of a most worthy dog-friend of my youth, well remembered by many now living who knew Greenwich Hospital some thirty or five-and-thirty years ago.

At that time there lived there a dog-pensioner called Hardy, a large brown Irish retriever. He was so named by Sir Thomas Hardy, when Governor (Nelson's Hardy), who at the same time constituted him a pensioner, at the rate of one penny per diem, for that he had one day saved a life from drowning just opposite the hospital.

Till that time he was a poor stranger and
vagrant dog—friendless.   But thenceforward
he lived in the hospital, and *spent his pension
himself* at the butcher's shop, as he did also
many another coin given to him by numerous
friends.   Many is the halfpenny which, as a
child, I gave Hardy, that I might see him
buy his own meat—which he did with judgment, and a due regard to value.   When a
penny was given to him, he would, on
arriving at the shop, place it on the counter
and rest his nose or paw upon it until he
received *two halfpennyworths*, nor would any
persuasion induce him to give up the coin
for the usual smaller allowance.   I was a
young child at the time, but I had a great
veneration for Hardy, and remember him
well, but lest my juvenile memory might
have been in fault, I have, before writing
this letter, compared my recollections with
those of my elders, who, as grown people,
knew Hardy for many years, and confirm
all the above facts.   There, indeed, was the
right dog in the right place.   Peace to his
shade!   J. D. C.

[*Feb.* 7, 1885.]

HAVE you room for one more dog story,
which resembles one lately reported in a
French journal ? A few years since I was
sitting inside the door of a shop to escape
from the rain while waiting for a trap to take
me to the railway station in the old Etruscan
city of Ferentino. Presently an ill-bred dog
of the pointer kind came and sat down in
front of me, looking up in my face, and
wagging his tail to attract my attention.
"What does that dog want ?" I asked of a
bystander. "Signore," he answered, "he
wants you to give him a soldo to go and
buy you a cigar with." I gave the dog the
coin, and he presently returned, bringing a
cigar, which he held crossways in his mouth
until I took it from him. Sent again and
again, he brought me three or four more
cigars from the tobacco-shop. At length the
dog's demeanour changed, and he gave vent
to his impatience by two or three low whines.
"What does he want now ?" I asked. "He
wants you to give him two soldi to go to the
baker's and buy bread for himself." I gave

him a two-soldo piece, and in a few minutes the dog returned with a small loaf of bread, which he laid at my feet, at the same time gazing wistfully in my face. "He won't take it until you give him leave," said another bystander. I gave the requisite permission, and the dear animal seized the loaf and disappeared with it in his mouth, and did not again make his appearance before I left the city. "He always does like this," said the standers-by, "whenever he sees a stranger in Ferentino."

GREVILLE I. CHESTER.

CAUTIOUS DOGS.

INTELLIGENT SUSPICION IN A DOG.

[*July* 7, 1888.]

THE following instance of dog instinct (or reasoning ?) will, I think, interest some of your readers.   About a fortnight ago, while crossing the Albula Pass, our driver stopped for a few moments at the little restaurant on the highest point of it.   A rough kind of herdsman's dog, of no particular breed, I suppose, came out and sat down by the carriage and looked up at us.   We happened to have a few Marie biscuits in the carriage, so I threw half of one out to him.   I suppose he had no experience in Huntley and Palmer's make, for he looked at and smelled it carefully, and then declined to eat it, but again looked up at me.   I then took the remaining half, bit off and ate a little bit of it, and then threw over the rest to him. This time he ate it at once, then turned and ate the first piece, which he had before refused, and at once came and asked for more, which I had great pleasure in giving him. I may add that I have several times tried a similar experiment with more pampered dogs at home, but have never succeeded with it. Whether this arises from the latter knowing, in most cases, from experience what they like and what they do not like, or, as I am rather inclined to think, from the superior intelligence of this Alpine dog, who really reasoned that what I could eat he could, I leave your readers to decide for themselves.

G. W. C.

AN ALPINE DOG.

[*July* 21, 1888.]

I DO not think that it was superior intelligence in the Alpine dog over other intelligent dogs which induced him to wait to eat the biscuit till he had seen the giver eat some of it. We have a very sagacious little Highland terrier, and he in the same manner often refuses a new kind of biscuit or cake until he has seen me bite off a small piece and eat it, and then he will do the same. I have also found our boarhound distrusting food occasionally, and declining to take it from his bowl until I have given him some with my hand. Then he seems to feel that it is all right, and comes down from his bench and eats it. This perhaps is not exactly the same, but it is still a phase of a dog's distrust of unaccustomed food, and his reasoning power respecting it. This wonderful reasoning power any one accustomed to dogs soon discovers.

J. B. G.

DOGS AND LANGUAGE.

DO DOGS UNDERSTAND OUR LANGUAGE?

[*Aug.* 4, 1883.]

I THINK the question has been mooted in your columns as to whether dogs sometimes understand our language. A circumstance that has just occurred leads me to think that it does happen, where they are highly organised and living much with their owners. While our family party were sitting over dessert, a cork jumped from an apollinaris-water bottle on the sideboard. I took no notice at first, but after the conversation was ended, I got up and looked about for a few minutes, soon giving up the search. My brother asked what I was looking for, and I answered. I had no sooner sat down than our little dog crept from behind a piece of furniture, where she was reposing on the end of a rug, and went straight up to the cork, looking up at me and pointing to it with her nose. It was near me, but the shadow thrown by the table prevented my seeing it.

She is a very nervous little fox-terrier, a most "comfort-loving animal," and spends her life with one or the other of us on my sofa, when her master is out, but hearing his voice at a great distance, and always attending to it.

ANYTHING BUT A DOG-FANCIER.

HOW OUR MEANING IS CONVEYED TO ANIMALS.

[*Aug.* 11, 1883.]

THE following anecdote may interest some of your readers :—Some years ago, when starting for a foreign tour, I entrusted my little Scotch terrier, Pixie, to the care of my brother, who lived about three miles distant from my house. I was away for six weeks, during the whole of which time Pixie remained contentedly at his new abode. The day, however, before I returned, my brother mentioned in the dog's hearing that I was expected back the next day. Thereupon, the dog started off, and was found by me at my bedroom door the next morning, he having been seen waiting outside the house early in the morning when the servants got up, and been admitted by them. Pixie is still alive and flourishing, and readily lends himself to experiments, which, however, yield no very definite result. He certainly seems to understand as much of our meaning as it concerns his own comfort to understand, but how he does it I cannot quite determine. I should be sorry to affirm, clever as he is, that he understands French and German, yet it is certainly a fact that he will fall back just as readily if I say "Zurück!' as if I say " To heel! " and advance to the sound " En avant! " as well as to " Hold up! " As in both cases I am careful to avoid any elucidatory gesture or special tone of voice, I am inclined to think that there must be here a species of direct thought transference. At the same time, I am bound to add that without the spoken word I am unable to convey the slightest meaning to him. This, however, may be due to what I believe to be a fact, that it is almost impossible without word or gesture to formulate the will with any distinctness. If this theory be correct, the verbal sounds used would convey the speaker's meaning, not in virtue of the precise sounds themselves, but of the intention put into them by the speaker. I should be glad to know if the experience of others tends to confirm this theory, which I do not remember to have seen suggested before.

A. EUBULE-EVANS.

[*Aug.* 18, 1883.]

I BEG to contribute another anecdote on the subject of how our meaning is conveyed to animals. When I was in Norway with my husband, a dog belonging to the people of the house went with us in all our walks. One day a strange dog joined us, and seemed to wish to get up a fight with our dog, Fechter, who for protection kept almost under our feet; my husband said several times, "Go on, Fechter," in English, which he immediately did, but soon came back again. At last we succeeded in driving the strange dog away, but he soon returned. Then my husband said without any alteration of tone or gesture that I was aware of, "Drive that dog away, Fechter." He immediately rushed at him, and we saw no more of our troubler. I have long thought that dogs do understand, not "the precise sounds themselves, but the intention put into them by the speaker."

An Observer of Animals.

ANIMAL INTELLIGENCE.

[Aug. 18, 1883.]

Perhaps I should have said the "Intelligence of Animals," but my meaning, in relation to the interesting correspondence in your columns, is no doubt clear. The whole question seems to me to lie in the proverbial nutshell, and to be solvable by the proverbial common sense. Dogs' hearing is undoubtedly very keen and accurate, and even subtle; and dogs have also the power of putting this and that together in a marvellously shrewd and almost rational fashion. They cannot understand sentences, but they get hold of words, *i.e.*, sounds, and keep them pigeon-holed in their memory. I might as well argue moral principle from the fact that my dog Karl, like scores of other dogs, will hold a piece of biscuit on his nose so long as I say "trust," and will when I say "paid for" gaily toss his head and catch the biscuit in his honest mouth, as argue that because he finds eleven tennis-balls among the shrubs in five minutes, when I say, "We can't find them at all, Karl; do go and find them, good dog, will you? Find the balls, old fellow"— therefore he understands my sentence. He simply grasps the words "find" and "balls," sees the game at a standstill, and reasons out our needs and his responsibilities, quickened by the expectation of pattings on the head, pettings, and pieces of biscuit. It is remarkable that if I try to delude him by uttering "base coin" in the shape of words just like the real words, as, for example, if I say "Jacob" instead of "paid for," he makes no mistake, but refuses the morsel, however delicate, till it *is* "paid for."

Prominent nouns, participles, verbs, &c., make up the *lingua franca* that so beautifully

links together men and dogs, and now and
then men and horses, their intelligence being
quickened by their dumbness, as is that of
deaf and dumb men and women, whose other
faculties become so keenly intensified, and
who put this and that together so much more
quickly than do we who have all our faculties.
There are of course " Admiral Crichtons "
among dogs, as there are among men, but
the difference between dog and dog will
generally, I think, be traceable more to
human training than to born capacity.   The
yearning look which Karl gives when (told
to "speak") he gives forth his voice in
response, is sometimes piteously like " Oh,
that I could really tell all I feel!"   He is
like, and all dogs of average intelligence are
like, the Frenchman I met yesterday on the
beach at Hastings, who wanted to know
whether he could reach Ramsgate on foot
before nightfall, and how far it was, and who,
as I only know a few French words, and am
utterly unable to speak or understand sentences, was obliged to make me understand
his wants by a few nouns such as everybody

knows, and by causing me to put this and that together. There is of course the vital defect in the parallel that I could learn to understand French, and the dog could never learn to understand sentences; but as so many parallels have vital defects of some kind, even down to that historic self-drawn parallel between Alexander and the robber, we may well say, whether we be men or dogs, "Let me reflect." Dogs do undoubtedly reflect, and reason, and remember; and they never forget their "grammar," as schoolboys do. Instinct, like chance, is only a name expressing fitly enough our own ignorance. Did not Luther and Wesley believe in the resurrection of animals?

S. B. JAMES.

[*Aug.* 25, 1883.]

A LITTLE illustration of canine intelligence shown by my collie, Dido, may be added to those which have lately appeared in the *Spectator.* The dog was lying on the floor in a room in which I was preparing to go out. An old servant was present, and when

I had given her directions about an errand on which she was going, I said, "You will take Dido with you?" She assented, and the dog directly got up to follow her down-stairs. I then remembered that I should want a cab, so I asked the servant to send one, and not to leave the house till I rang the bell. On her leaving the room, Dido resumed her quiet attitude on the floor, with her nose to the carpet. In rather less than ten minutes I rang the bell, and the dog at once sprang up and ran downstairs to join her companion. I had not spoken a word after asking the servant to wait for the bell. Was this word-reading, or voice-reading, or thought-reading.

S. E. DE MORGAN.

ANIMALS AND LANGUAGE.

[*Sept.* 1, 1883.]

I CAN match Mrs. De Morgan's pretty story of her Dido. A wise old dog with whom I have the privilege to associate was, two or three days ago, lying asleep in her basket by the fire. I entered the room with my hat on, and invited her to join me in a walk; but, after looking up at me for a moment, as canine politeness required, she dropped back among her cushions, obviously replying, "Thank you very much, but I prefer repose." Thereupon I observed, in a clear voice, "I am *not* going on the road [a promenade disliked by the dogs, because the walls on either side restrict the spirit of scientific research]; I am going up the mountain." Instantly my little friend jumped up, shook her ears, and, with a cheerful bark, announced herself as ready to join the party.

Beyond doubt or question, Colleen had either understood the word "road," or the word "mountain," or both, and determined her proceedings accordingly. Nothing in my action showed, or could show, the meaning of my words.

If any of your readers who have resided for some weeks or months in a country where a language is spoken entirely foreign to their own—say, Arabic, or Basque, or Welsh—will recall of how many words they insensibly learn the meaning without asking it, and merely by hearing them always used in certain relations, they will have, I think, a fair measure of the extent and nature of a dog's knowledge of the language of his masters. My dog has lived fewer years in the world than I have passed in Wales, but he knows just about as much English as I know Welsh, and has acquired it just in the same way.

F. P. C.

TEACHING DOGS A METHOD OF COMMUNICATION.

[*Dec.* 29, 1883.]

MR. DARWIN'S "Notes on Instinct," recently published by my friend, Mr. Romanes, have again called attention to the interesting subject of instinct in animals.

Miss Martineau once remarked that, considering how long we have lived in close association with animals, it is astonishing how little we know about them, and especially about their mental condition. This applies with especial force to our domestic animals, and, above all, of course, to dogs.

I believe that it arises very much from the fact that hitherto we have tried to teach animals, rather than to learn from them— to convey our ideas to them, rather than to devise any language, or code of signals, by means of which they might communicate theirs to us. No doubt the former process is interesting and instructive, but it does not carry us very far.

Under these circumstances it has occurred to me whether some such system as that followed with deaf mutes, and especially by Dr. Howe with Laura Bridgman, might not prove very instructive if adapted to the case of dogs. Accordingly I prepared some pieces of stout cardboard, and printed on each in legible letters a word, such as " food," "bone," "out," &c. I then began training a black poodle, Van by name, kindly given me by my friend, Mr. Nickalls.

I commenced by giving the dog food in a saucer, over which I laid the card on which was the word " food," placing also by the side an empty saucer, covered by a plain card. Van soon learnt to distinguish between

the two, and the next stage was to teach
him to bring me the card ; this he now does,
and hands it to me quite prettily, and I then
give him a bone, or a little food, or take him
out, according to the card brought. He
still brings sometimes a plain card, in which
case I point out his error, and he then takes
it back and changes it. This, however, does
not often happen. Yesterday morning, for
instance, he brought me the card with
"food" on it nine times in succession,
selecting it from among other plain cards,
though I changed the relative position every
time. No one who sees him can doubt that
he understands the act of bringing the card
with the word "food" on it, as a request for
something to eat, and that he distinguishes
between it and a plain card. I also believe
that he distinguishes, for instance, between
the card with the word "food" on it and the
card with "out" on it.

This, then, seems to open up a method
which may be carried much further, for it is
obvious that the cards may be multiplied,
and the dog thus enabled to communicate

freely with us. I have as yet, I know, made only a very small beginning, and hope to carry the experiment much further, but my object in troubling you with this letter is ʰwofold. In the first place, I trust that some of your readers may be able and willing to suggest extensions or improvements of the idea. Secondly, my spare time is small, and liable to many interruptions; and animals also, we know, differ greatly from one another. Now, many of your readers have favourite dogs, and I would express a hope that some of them may be disposed to study them in the manner indicated. The observations, even though negative, would be interesting; but I confess I hope that some positive results might follow, which would enable us to obtain a more correct insight into the minds of animals than we have yet acquired.

JOHN LUBBOCK.

COMMUNICATION WITH ANIMALS.

[*April* 12, 1884.]

You did me the honour, some weeks ago, to insert a letter of mine, containing suggestions as to a method of studying the psychology of animals and a short account of a beginning I had myself made in that direction.

This letter has elicited various replies and suggestions which you will perhaps allow me to answer, and I may also take the opportunity of stating the progress which my dog Van has made, although, owing greatly, no doubt, to my frequent absences from home and the little time I can devote to him, this has not been so rapid as I doubt not would otherwise have been the case. Perhaps I may just repeat that the essence of my idea was to have various words, such as "food," "bone," "water," "out," &c., printed on pieces of card-board, and, after some preliminary training, to give the dog anything for which he asked by bringing a card. I use pieces of cardboard about ten inches long and three inches high, placing a number of them on the floor side by side, so that the dog has several cards to select from, each bearing a different word.

One correspondent has suggested that it would be better to use variously coloured cards. This might, no doubt, render the first steps rather more easy, but, on the other hand, any temporary advantage gained would be at the expense of subsequent difficulty, since the pupil would very likely begin by associating the object with the colour, rather than with the letters. He would, therefore, as is too often the case with our own children, have the unnecessary labour of unlearning some of his first lessons. At the same time, the experiment would have an interest as a test of the colour-sense in dogs.

Another suggestion has been that, instead of words, pictorial representations should be placed on the cards. This, however, could only be done with material objects, such as " food," " bone," " water," &c., and would not be applicable to such words as " out," " pet me," &c. ; nor even as regards the former class do I see that it would present any substantial advantage.

Again, it has been suggested that Van is led by scent rather than by sight. He has, no doubt, an excellent nose, but in this case he is certainly guided by the eye. The cards are all handled by us, and must emit very nearly the same odour. I do not, however, rely on this, but have in use a number of cards bearing the same word. When, for instance, he has brought a card with "food" on it, we do not put down the same identical card, but another with the same word; when he has brought that, a third is put down, and so on. For a single meal, therefore, eight or ten cards will have been used, and it seems clear, therefore, that in selecting them Van must be guided by the letters.

When I last wrote I had satisfied myself that he had learnt to regard the bringing of a card as a request, and that he could distinguish a card with the word "food" on it from a plain one, while I believed that he could distinguish between a card with "food" on it and one with "out' on it.

I have now no doubt that he can distinguish between different words. For instance, when he is hungry he will bring a "food" card time after time, until he has had enough, and then he lies down quietly for a nap. Again, when I am going for a walk, and invite him to come, he gladly responds by picking up the "out" card, and running triumphantly with it before me to the front door. In the same way he knows the "bone" card quite well. As regards water (which I spell phonetically, so as not to confuse him unnecessarily), I keep a card always on the floor in my dressing-room, and whenever he is thirsty he goes off there, without any suggestion from me, and brings the card with perfect gravity. At the same time he is fond of a game, and if he is playful or excited will occasionally run about with any card. If through inadvertence he brings a card for something he does not want, when the corresponding object is shown him, he seizes the card, takes it back again, and fetches the right one. No one who has seen him look along a row of cards, and select the right one, can, I think, doubt that in bringing a card he feels that he is making a request, and that he can not only perfectly distinguish between one word and another, but also associates the word and the object.

I do not for a moment say that Van thus

shows more intelligence than has been recorded in the case of other dogs; that is not
my point, but it does seem to me that this
method of instruction opens out a means by
which dogs and other animals may be enabled
to communicate with us more satisfactorily
than hitherto.   I am   still continuing my
observations, and am now considering the
best mode of testing him in very simple
arithmetic, but I wish I could induce others
to co-operate, for I feel satisfied that the
system would well repay more time and
attention than I am myself able to give.

JOHN LUBBOCK.

INSTINCT OF LOCALITY IN DOGS.

[March 4, 1893.]

A CAT carried a hundred miles in a basket, a dog taken, perhaps, five hundred miles by rail, in a few days may have found their way back to the starting-point. So we have often been told, and, no doubt, the thing has happened. We have been astonished at the wonderful intelligence displayed. Magic, I should call it. Last week I heard of a captain who sailed from Aberdeen to Arbroath. He left behind him a dog which, according to the story, had never been in Arbroath, but when he arrived there the dog was waiting on the quay. I was expected to believe that the dog had known his master's destination, and been able to inquire the way overland to Arbroath. Truly marvellous! But, really, it is time to inquire more carefully as to what these stories do mean ; we must cease to ascribe our intelligence to animals, and learn that it is we that often possess their instinct. A cat on a farm will wander many miles in search of prey, and will therefore be well acquainted with the country for many miles round. It is taken fifty miles away. Again it wanders, and comes across a bit of country it knew before. What more natural than that it should go to its old home? Carrier-pigeons are taught " homing " by taking them gradually longer flights from home, so that they may learn the look of the country. We cannot always discover that a dog actually was acquainted with the route by which it wanders home; but it is quite absurd to imagine, as most people at once do, that it was a perfect stranger to the lay of the land. To find our way a second time over ground we have once trod is scarcely intelligence; we can only call it instinct, though the word does not in the least explain the process. Two years ago I first visited Douglas, in the Isle of Man. I reached the station at 11 p.m.; I was guided to a house a mile through the town. I scarcely paid any attention to the route, yet next morning I found my way by the same route to the station, walking with my head bent, deeply thinking all the time about other things than the way. I have the instinct of locality. Most people going into a dark room that they know are by mus-cular sense guided exactly to the very spot they wish ; so people who have the instinct of locality may wander over a moor exactly to the place they wish to reach without thinking of where they go. There may be no mental exercise connected with this. I have known a lady of great intelligence who would lose her way within half-a-mile of the house she had lived in forty years. This feeling about place belongs to that part of us that we have in common with the lower creatures. We need not postulate that the animals ever show signs of possessing our intelligence ; they possess, in common with us, what is not intelligence, but instinct.

A. J. MACKINTOSH.

[*Sept.* 24, 1892.]

WILL you allow me to record in the *Spectator* "another dog story"? It is one that testifies, for the thousandth time, to canine sagacity, and, as we are still in the silly season, which has this year in particular been so very prolific in human follies, it may be of special interest to learn some clever doings on the part of beasts. Quite recently a Westphalian squire travelled by rail from Lüxen to Wesel, on the Rhine, for the purpose of enjoying some hunting, and took with him his favourite hound. The hunting party was to have started on a Sunday morning at nine o'clock, but, to the squire's great disappointment, his sporting dog could nowhere be discovered. Disconsolate, he arrived on the following Monday afternoon at his house, and, to his great delight, he was greeted there with exuberant joy by his dog. The latter, who had never made the journey from Lüxen to Wesel, had simply run home, thus clearing a distance of eighty English miles through an unknown country. Why the sporting dog should have declined to join the hunt is, perhaps, a greater mystery than the fact of his returning home without any other guidance than his sagacious instinct. Possibly he was a Sabbatarian, and objected

to imitate his master's wicked example.   So,
Sunday papers, please copy !

EIN THIERFREUND.

[*Sept.* 8, 1894.]

MAY I be allowed to offer to your readers
yet another instance of the faithfulness and
sagacity of our friend the dog ?   The anecdote comes from a distinguished naval
officer, and is best given in his own words :
" This is what happened to a spaniel of mine.
It was given to our children as a puppy
about three or four months old, and we have
had it about five or six months, making it
about ten months old.   It was born about
three miles from here, at Hertford, and has
never been anywhere but from one home to
the other.   When the time came for break-
ing him in for shooting purposes, I sent him
to a keeper at Leighton-Buzzard, and, to
insure a safe arrival, sent the dog with my
man-servant to the train here, and thence to
King's Cross.   He walked with the dog to
Euston Station, turned him over to the guard
of the 12.15 train and the animal duly

arrived at Leighton-Buzzard at 1.30, and was there met by the keeper and taken to his home about three miles off. That was on the Friday. On the following Tuesday, the dog having been with him three full days, he took him out in the morning with his gun, and at eight o'clock on Wednesday morning (that being the following day) the dog appeared here, rather dirty, and looking as if he had travelled some distance, which he undoubtedly had. There is no doubt that this puppy of ten months old was sent away, certainly forty or fifty miles as the crow flies, and that he returned here in a day. How he did it no one can say, but it is nevertheless a fact. It would be interesting to know his route and to trace his adventures." This anecdote is the more remarkable in consequence of the extreme youth of the dog, and particularly as he belongs to a breed of sporting dogs which are not generally considered to rank among the most intelligent of the species.

F. H. SUCKLING.

[*Sept.* 15, 1894.]

THE " True Story of a Dog," in the *Spectator* of September 8th, may be matched, possibly explained, by a similar occurrence. I had bought a Spanish poodle pup of an Irishman who assured me, " Indade, sir, an' the dog knows all my childer do, only he can't talk." He shut doors, opened those with thumb-latches, and rushed upstairs and waked his mistress at words of command. One day we were starting to drive to our former home in the city, six miles distant, but the dog was refused his usual place in the carriage, and shut up in the house. When we arrived, to our astonishment we found him waiting for us on the doorstep ! We could not conceive how he got there, but upon inquiry found that he had got out, gone to the station, in some way entered the train, hid under a seat, and on arrival in the city threaded his way a mile through the streets, and was found quietly awaiting our arrival.

R. P. S.

[*May* 3, 1884.]

How do we know that in inviting dogs to
the use of words Sir John Lubbock is
developing their intelligence? Are we sure
that he is not asking them to descend to
a lower level than their own, in teaching
them to communicate with us through our
proper forms of speech, unnecessary to
them? I can vouch for the truth of the
following story. A young keeper, living
about twelve miles east of Winchester, on
leaving his situation gave away a fox-terrier,
which had been his constant companion for
some months; he then took another place
in the north of Hampshire, near the borders
of Berkshire, in a part of the country to
which he had never been. The new owner
of the dog took her with him to a village
in Sussex; before she had been there long
she disappeared, and after a short time found
her old master in the woods at his new home.
As I have said before, he had never been
there before, neither had she. Rather un-
gratefully, he again gave the dog away, this
time to a man living some way north of

Berkshire; she came back to him in a few days, and, I am happy to say. is now to be allowed to stay with the master of her choice. Can such a nature need to be taught our clumsy language.

A H. WILLIAMS.

[*Feb.* 16, 1895.]

As I see that you have published some interesting anecdotes about dogs, I send you the two following, which perhaps you may think worth inserting.

In 1873 we came to live in England, after a residence upon the Continent, bringing with us a Swiss terrier of doubtful breed but of marked sagacity, called Tan. One day, shortly after reaching the new home from Switzerland, the dog was lost under the following circumstances:—We had driven to a station eight miles off—East Harling— to meet a friend. As the friend got out of the railway carriage the dog got in without being noticed and the train proceeded on its way. At the next station—Eccles Road —the dog's barking attracted the attention of the station-master, who opened the carriage door, and the dog jumped out. The station-master and the dog were perfect strangers. He and a porter tried to lock up the dog, but he flew viciously at any one who attempted to touch him, although he was not above accepting food. For the next three days his behaviour was decidedly methodical; starting from the station in the morning, he came back dejected and tired at night. At last, on the evening of the third day, he reached home, some nine miles away, along roads which he had not before travelled, a sorry object and decidedly the worse for wear; after some food he slept for twenty-four hours straight off.

Anecdote number two. One day a handsome black, smooth-haired retriever puppy was given to us, whom we named Neptune. The terrier Tan greatly resented having this new companion thrust upon him, and became very jealous of him. Being small, he was unable to tackle so large a dog, but sagacity accomplished what strength could not. Tan disappeared for two days. One evening, hearing a tremendous commotion in the yard, we rushed out to find a huge dog of the St. Bernard species inflicting a severe castigation upon poor Nep, Tan meanwhile looking on, complacently wagging his tail. Both Tan and his companion then disappeared for two more days, after which Tan reappeared alone, apparently in an equable frame of mind, and satisfied that he had had his revenge. We never discovered where the large dog came from. I can attest the truth of the two stories.

CECIL DOWNTON.

RAILWAY DOGS.

[*July* 30, 1887.]

YOUR dog-loving readers may be interested to hear that there is (or was till lately) in South Africa a rival to the well-known Travelling Jack, of Brighton line fame, after whom, indeed, he has been nicknamed by his acquaintance.

I was introduced to him eighteen months ago, on board the *Norham Castle*, on a voyage from Cape Town to England—a voyage which this distinguished Colonial traveller was making much against his will. He was a black-and-tan terrier with a white chest, whose intellect had therefore probably been improved by a dash of mongrelism, and I was told that he belonged to a gentleman connected with the railway department living at Port Elizabeth. It appears that it was Mr. Jack's habit frequently to embark all by himself on board the mail steamer leaving that place on Saturday afternoon, and make the trip round the coast to Cape Town, arriving there on Monday morning. Where he " put up " I do not know, but he used to stay there until Wednesday evening, when he would calmly walk into the station, take his place in the train, and return to Port Elizabeth in that way, thus completing his " circular tour " by a railway journey of about eight hundred miles.

He was well known by the officers and sailors of the *Norham*, and her commander, Captain Alexander Winchester (who can vouch for these facts), told me that, as the dog seemed fond of the sea, he had determined to give him a long voyage for a change, and had kept him shut up on board during the ship's stay at Cape Town.

Jack was evidently very uneasy at being taken on beyond his usual port, and he was on the point of slipping into a boat for the shore at Madeira, probably with a view of returning to the Cape by the next steamer, when I called the captain's attention to him, and he was promptly shut up again. I said good-bye to him at Plymouth, and hope he found his way home safely on the return voyage.

Ex-Colonist.

[*June* 23, 1894.]

I HAVE read with much interest the stories in the *Spectator* of the sagacity of animals. The following, I think, is worth recording:—The chief-engineer of the Midland and South-Western Junction Railway, Mr. J. R. Shopland, C.E., has a spaniel that frequently accompanies him or his sons to their office. On Saturday last this dog went to Marlborough from Swindon by train with one of Mr. Shopland's clerks, and walked with him to Savernake Forest. Suddenly the dog was missing. The creature had gone back to the station at Marlborough and taken a seat in a second-class compartment. The dog defied the efforts of the railway officials to dislodge him. When the train reached Swindon he came out of the carriage and walked quietly to his master's residence.

SAMUEL SNELL.

[*March* 30, 1895.]

I WAS witness the other day of what I had only heard of before—a dog travelling by rail on his own account. I got into the train at Uxbridge Road, and, the compartment being vacant, took up the seat which I now prefer—the corner seat at the entrance with the back to the engine. Presently a whole crowd of ladies got in, and with them a dog, which I supposed to belong to them. All the ladies except one got out at Addison Road, and then the dog slunk across the carriage to just under my seat. I asked my remaining fellow-passenger whether the dog was hers; she said "No." No one got in before she herself got out at South Kensington, where the dog remained perfectly quiet, but at Sloane Square a man was let in, and out rushed the dog, the door actually grazing his sides. Had he not taken up the precise place he did, he must have been shut in or crushed. "That dog is a stowaway," I observed to the porter who had opened the door. "I suppose he is," the man answered. The dog was making the best of his way to the stairs. Clearly the dog meant to get out at that particular station (he had had ample opportunity of getting out both at Addison Road

and South Kensington), and had, as soon
as he could, taken up the best position for
doing so.   How did he recognise the Sloane
Square Station, for he had had only those
two opportunities of glancing out?   It seems
to me it could only have been by counting
the stations, in which case he must be able
to reckon up to five.   The dog was a very
ordinary London cur, white and tan, of a
greatly mixed Scotch terrier stock, the long
muzzle showing a greyhound cross.   He
was thin, and apparently conscious of break-
ing the law, hiding out of sight, and slinking
along with his tail between his legs, and
altogether not worth stealing.   I suppose
that he had been transferred to a new home
which had proved uncongenial, and was
slipping away, in fear and trembling, to his
old quarters.

J M. L.

EMOTION AND SENTIMENT IN DOGS.

A DOG'S REMORSE.

[*Sept.* 1, 1883.]

A REMARKABLE instance of the effect that can be produced upon a dog by the human voice was related to me yesterday. Some of your correspondents would consider it confirmatory of their notion that dogs have mind enough to understand words; but I myself rather believe that the sound of the voice acts upon the *feelings* of dumb animals just as instrumental music acts upon us. The story is as follows :—A clergyman had for a long time a dog, and no other domestic animal. He and his servant made a great pet of the dog. At last, however, the clergyman took to keeping a few fowls, and the servant fed them. The dog showed himself very jealous and out of humour at this, and when Sunday came round, and he was left alone, he took the opportunity to *kill and bury* two hens.

A claw half-uncovered betrayed what he had done. His master did not beat him, but took hold of him, and *talked* to him, most bitterly, most severely. "You've been guilty of the sin of murder, sir,—and on the Sabbath day, too; and you, a clergyman's dog, taking a mean advantage of my absence!" &c. He talked on and on for a long time, in the same serious and reproachful strain. Early the next morning the master had to leave home for a day or so; and he did so without speaking a word of kindness to the dog, because he said he wished him to feel himself in disgrace. On his return, the first thing he was told was, "The dog is dead. He never ate nor drank after you had spoken to him; he just lay and pined away, and he died an hour ago."

L. G. GILLUM.

A CONSCIENCE-STRICKEN DOG.

[*Feb.* 1, 1879.]

You have frequently published letters containing stories bearing on the question of the moral nature and the future of the lower animals. I venture to send you some facts about a dog, narrated to me by a lady, whose name and address I enclose for your own satisfaction, and at my request written down by her as follows—

" A young fox-terrier, about eight months old, took a great fancy to a small brush, of Indian workmanship, lying on the drawing-room table. It had been punished more than once for jumping on the table and taking it. On one occasion, the little dog was left alone in the room accidentally. On my return, it jumped to greet me as usual, and I said, ' Have you been a good little dog while you have been left alone ? ' Immediately it put its tail between its legs and slunk off into an adjoining room, and brought back the little brush in its mouth from where it had hidden it.

"I was much struck with what appeared to me a remarkable instance of a dog possessing a conscience, and a few months afterwards, finding it again alone in the room, I asked the same question, while patting it. At once I saw it had been up to some mischief, for with the same look of shame it walked slowly to one of the windows, where it lay down, with its nose pointing to a letter bitten and torn into shreds. On a third occasion, it showed me where it had strewn a number of little tickets about the floor, for doing which it had been reproved previously. I cannot account for these facts, except by supposing the dog must have a conscience."

The conduct of this dog seems to me, sir, to exhibit something different from fear of punishment, viz., a sense of shame, a remorse, a desire to confess his fault, and even to expiate it by punishment, in order to feel the guilt no longer. He rather sought punishment, than feared it.

TH. HILL.

A DOG'S AFFECTION.

[April 24, 1875.]

I SAW an anecdote in your paper the other week illustrative of the sagacity of a dog. Kindly allow me to place upon record, as a kind of a companion picture, an anecdote showing the *affection* of one of the canine species—a fine young retriever. For some weeks I have been staying away from my house in the country, where is the fine young retriever in question. Well, last week the household missed him for hours, and began to think he was lost. Nothing of the kind, however. The servant, happening to go up to my bedroom, found him with his head resting on my pillow, moaning heavily, and it was only with great difficulty that she could drive him away. Surely it is incidents such as these that have made so many great men rail against humanity and uphold their dog!

WILL WILLIAMS.

AFFECTION.

[*Sept.* 15, 1894.]

As you sometimes admit anecdotes of animals into the *Spectator*, perhaps you may consider the following fact worthy of record. In a hotel where I am staying, being distressed by the cry of anguish of a dog occasionally, I inquired the cause, and was told that whenever he happens to be in the hall when luggage is brought down to go in the omnibus, he utters these bitter cries, and has to be removed. His master left him here many months ago, and the supposition is that the sight of the luggage and omnibus recalls his loss; and is another instance of the faithful affection of these half-human creatures.

J. K.

SYMPATHY IN A DOG.

[*July* 30, 1892.]

THE article, "Animals in Sickness," in the *Spectator* of July 23rd, has reminded me of the following anecdote, which was told to me some years ago by a butcher residing at Brodick, in the Isle of Arran. He told me that he had had two collie dogs at the same time, one old and the other young. The old dog became useless through age, and was drowned in the sea at Brodick. A few days afterwards, its body was washed ashore, and it was discovered by the young dog, who was seen immediately to go to the butcher's shop and take away a piece of meat and lay it at the dead dog's mouth. The young dog evidently thought that the meat would revive his old comrade, and thereby showed remarkable sympathy in aid of, to him, the apparent "weak."

DAVID HANNAY.

A DOG'S HUMANITY.

[April 18, 1891.]

POSSIBLY it is from an excess of the "maudlin
sentimentality" of which physiologists com-
plain in those who protest against cruelty to
animals, that I find it almost painful to read
such pathetic stories of dogs as the one
given by Miss Cobbe in the *Spectator* of
April 11th ; for they tell of such intelligence
and devotion, that, remembering the in-
human way in which our poor dogs are too
often treated, we feel it would be almost
better if they lacked these human qualities.

The following is an anecdote of the same
kind, that ever since I heard it, I have been
intending to send it to the *Spectator*. The
servant-man of one of my friends took a
kitten to a pond with the intention of drowning it. His master's dog was with him, and
when the kitten was thrown into the water,
the dog sprang in and brought it back safely
to land. A second time the man threw it in,
and again the dog rescued it ; and when for
the third time the man tried to drown it, the

dog, as resolute to save the little helpless life as the man was to destroy it, swam with it to the other side of the pool, running all the way home with it, and safely depositing it before the kitchen fire ; and "ever after" they were inseparable, sharing even the same bed !

When not long ago I came across the noble sentiment that "hecatombs of brutes should be tortured, if man thereby could be saved one pang," I found myself dimly wondering what constituted a "brute." Certainly, in the incident I have just given, the "brute" was not the dog !

S. W.

A CANINE MEMBER OF THE S.P.C.A.

[*June* 18, 1892.]

IF you think this little anecdote of canine friendliness worthy of the *Spectator*, will you insert it for me? Last week a sick dog took up its abode in the field behind our house, and after seeing the poor thing lying there for some time, I took it food and milk-and-water. The next day it was still there, and when I was going out to feed it, I saw that a small pug was running about it, so I took a whip out with me to drive it away. The pug planted itself between me and the sick dog, and barked at me savagely, but at last I drove it away, and again gave food and milk-and-water to my *protegé*. The little pug watched me for a few moments, and as soon as he felt quite assured that my intentions towards the sick dog were friendly, it ran to me wagging its tail, leapt up to my shoulder, and licked my face and hands, nor would it touch the water till the invalid had had all it wanted. I suppose that it was satisfied that its companion was in good hands, for it trotted happily away, and did not appear upon the scene again.

VIOLET DAVIES.

A DOG'S COURTESY.

[*Nov.* 29, 1890.]

In your article on Mr. Nettleship's pictures
of animals, you note the delicacy of a dog
that has been properly trained in the matter
of taking its food.   My little dog is not only
most dainty in that particular, but strictly
observes the courtesy, which is natural, not
taught, of not beginning his dinner (served
on white napery that is never soiled) until
his master begins his own.   No amount of
coaxing on the part of the ladies (they do
not wait) will induce him to eat if I am late :
he merely consents to have his muzzle taken
off, inspects his dinner, and then seeks his
master's room, where he waits to accompany
him in orderly fashion downstairs.

C. Harper.

CANINE JEALOUSY.

[*Dec.* 12; 1891.]

I AM not versed in dog-lore, and it may be that my love for the animal makes me an ill judge of the importance of the following story ; but a friend vouches for its truth, and to my mind it has its importance, not from its display of jealousy, but from the dog's deliberate acceptance of the undoubtedly changed condition, and the clearly metaphysical character of his motive.

The story is this. A young man had owned for some years a dog who was his constant companion. Recently the young man married, and moved with his bride and his dog into a house on the opposite side of the street from his father's house, his own former home. The dog was not happy, for the time and attention which had formerly been his was now given to the young wife. In many ways he showed his unhappiness and displeasure, in spite of the fact that the master tried to reconcile him and the bride to win him. One day when the master came home, his wife sat on his knee, while Jack was lying by the fire. He rose from his place, came over to the couple, and expressed his disapproval. "Why, Jack," said the master, "this is all right, she's a good girl," and as he spoke, he patted her arm. Jack looked up at him, turned away, and left the room. In a moment they heard a noise, and going into the hall, they found Jack dragging his bed downstairs. When he reached the front door, he whined to be let out, and when the door was opened, he dragged his bed down the steps, across the street to his old home, where he scratched for admittance. Since then he has never been back to his master, refusing all overtures.

CHAS. MORRIS ADDISON.

A JEALOUS DOG.

[*Jan.* 12, 1895.]

I was greatly interested in the story of the
generosity shown by a dog, as related in the
Spectator of January 5th, because of a similar
case within my own knowledge, and yet so
different, as to prove that the dispositions
of animals are as varied as those of human
beings.   A friend of mine had two fox-
terriers, inseparable companions, and both
equally devoted to their mistress.   On one
occasion, when the family had been away
from home for some time, and were returning, one of these pets, not being well, was
brought back with its mistress, while the
other was left to follow with the horses, &c.,
and did not arrive for three days.   On
entering the house, the dog had a very
sullen appearance, took no notice of any one,
but searched everywhere till he found his
companion ; then flew at his throat, and
would have killed him but for timely succour!
Could any human being have indulged in a
more rankling jealousy ?

A DOG THAT SCORNED TO BE JEALOUS.

[*Jan.* 5, 1895.]

THE following history of canine sympathy may interest your readers. I was once the happy owner of a large and beautiful bull-terrier, Rose, and at the same time of a still dearer, though less beautiful, little mongrel, Fan, both passionately attached to a member of my household, commonly called their best friend. A certain shawl belonging to this adored friend was especially sacred in Fan's eyes. She never allowed any one to touch it without remonstrance—Rose least of all—and when her best friend was in bed, it was Fan's custom to ensconce herself in her arms, and not to allow any dog, and only the most favoured of human beings, to approach without violent growlings, if not worse. Fan was a tiny grandmother who had long ruled the household; Rose, an inexperienced new-comer. One day, in a fit of youthful folly, Rose jumped over a gate and spiked herself badly, and was consigned for ten days to the care of the veterinary surgeon. On her return, she was cordially welcomed by Fan and myself; but when she rushed upstairs to the room of her best friend (then confined to her bed), my mind forboded mischief. We followed, and I opened the door. With one bound Rose flew into her best friend's arms, taking Fan's very own place, and was lost in a rapture of licking and being caressed. Fan flew after her, but to my amazement, instead of the fury I expected, it was to join with heart and tongue in the licking and caressing. She licked Rose as if she had been a long-lost puppy, instead of an intruder; and then, of her own accord, turned away, leaving Rose in possession, and took up a distant place on the foot of the bed, appealing to me with an almost human expression of mingled feelings—the heroic self-abnegation of new-born sympathy struggling with natural jealousy. The better feelings triumphed (not, of course, unsupported by human recognition and applause), till both dogs fell asleep in their strangely reversed positions. After this, there was a slight temporary failure in Fan's perhaps overstrained self-conquest; but on the next day but one she actually, for the first (and last) time in her life, made Rose welcome to a place beside her on the sacred shawl; where again they slept side by side like sisters. This, however, was the last gleam of the special sympathy called forth by Rose's troubles. From that day Fan decidedly and finally resumed her jealous occupation and guardianship of all sacred places and things, and maintained it energetically to her life's end.

C. E. S.

DOGS AND THE ARTS.

MUSIC AND DOGS.

[*Oct.* 24, 1891.]

DOGS, as well as horses, can recognise tunes. Many years ago a friend, during a short absence from our station on the Kurrumfooler, lent my sister a pet dog. Cissie was constantly in the room while playing and singing went on, without taking any notice ; but whenever the temporary mistress began singing one favourite song of the absent mistress's, the dog would jump on a chair by her side with evident pleasure.

O. H. G.

[*Oct.* 24, 1891.]

I HAVE read with much interest your correspondent's letter on the capability of animals to distinguish tunes. I had a small dog who, when first I got him, would have howled incessantly during singing. This, however, he was not allowed to do, except to one tune, which he soon knew and always joined in, not attempting to "sing" other songs. We tried every sort of experiment to see if he would recognise his own tune, which he invariably did, and would whine if the air was hummed quite quietly.

C. F. HARRISON.

[*Oct.* 24, 1891.]

ANENT "Orpheus at the Zoo," the following facts may interest you. Of two dogs of mine, one showed a great fondness for music. She (though usually my shadow) would always leave me to go to a room where a piano was being played, and the more she liked the music, the closer she crept to the player, even if a stranger to her. If, however, one began to play scales or exercises, she would get up, walk to the door, sit down, and, after waiting a bit, go away out of sight, but not out of hearing, for she soon appeared again on the resumption of music to her taste. On the other hand, mere "strumming" very quickly obliged her to go right away out of hearing. I confess that I have many times plagued the poor dog by thus sending her backwards and forwards. Her looks were often very comical. The other dog evidently hated music—would try to push a player from the piano, go out of hearing, and show other unmistakable signs of dislike. A band would draw one dog out to listen, while the other rushed away to hide. In one house the dog first mentioned had, for some reason or other, a particular objection to the room where the piano was, and never willingly stayed there. Music would bring her in, but only to sigh and moan, evidently in great pity for herself at being obliged to listen under such (to her) trying conditions. From these and other observations I am convinced that there is the musical dog as well as the unmusical, just as with human beings. D.

RECOGNITION OF LIKENESSES BY DOGS.

[*May* 5, 1894.]

IN the *Spectator* of April 21st there is an article on Apes, in which the following occurs : — " Monkeys, we believe, alone among animals can recognise the meaning of a picture." It may interest some of your readers to hear that certain other animals can also do this, two instances having come under my own observation. A cat belonging to a little girl I know was on the child's bed one morning, and made a spring at a picture of a thrush, about life-size, which was hanging near. The other case is that of a dog— a female Irish terrier—who is in the habit of running with her mistress's pony carriage. When she sees the pony being harnessed, she often shows her delight by jumping up at its head and barking. In a certain shop to which she sometimes goes with her mistress there is a picture of a horse hanging. The dog invariably behaves in exactly the same manner to this, jumping up and barking at it, thus showing unmistakably that she recognises its meaning.

JULIA ANDREWS.

May 19, 1894.

THE following instance bears on the subject discussed in the *Spectator* of May 5th. We had for a newcomer to our circle a little terrier dog. I was informed it had been seen in the library facing a large-sized portrait of myself, and barking furiously. I was somewhat sceptical until a day or two later I saw it repeat the performance. I have wondered whether it was because the dog thought it a good or bad representation of the original, and so was complimenting or otherwise the artist.

FRANK WRIGHT.

[*May* 19, 1894.]

APROPOS of the recognition of pictures by dogs (*Spectator*, May 5th), I think you may be interested in the two following facts which came under my notice a few years ago. A sagacious but quite uneducated old terrier came with his master to call for me, and coiled himself on the hearthrug while we talked. Turning himself round in the intervals of slumber, his eye caught an oil-painting just over his head (a life-size half-length of a gentleman). He immediately sat up, showed his teeth, and growled—not once, but continually—as both angry and mortified that neither eyes nor nose had given him notice of the arrival of a stranger! The next instance was similar, except that the chief actor was a young, intelligent collie, who, on the sudden discovery of a man looking at him from the wall, barked long and furiously. In both instances, after their excitement had subsided, I led the dogs to look at another picture similar in size, and also of a gentleman, but neither of them would take the smallest notice of it. I need only add that the picture which the dogs appreciated was painted by Sir Henry Raeburn—the other was not. Might not a few sagacious canine members be a useful addition to the Royal Academy Hanging Committee?

B. Thomson.

[*May* 26, 1894.]

MANY years ago I had a similar experience
to Mr. Frank Wright. A likeness of myself,
head and shoulders, drawn in chalk from a
photograph, and enlarged to nearly life size,
hung on the dining-room wall of a house I
then occupied. One evening my wife silently
called my attention to a young English
terrier, who had not been very long with
us, looking up at it very steadfastly. He
regarded it for about a minute in silence,
and at last broke out into a loud bark,
which I supposed to mean that in his
opinion the wall was not my proper place,
and that only an evil genius could have
set anything like me in such a position.

G.

[*June* 2, 1894.]

YOU were so good as to insert my little
account of the politeness of a parrot in the
Spectator, will you now allow me also to bear
witness to the recognition of a likeness by a
dog? Some time ago I was painting two
portraits in the country, and one day by

chance I placed the picture of my hostess on the ground. Immediately her old spaniel came and gazed intently at the face for several seconds. Then he smelt at the canvas, and, unsatisfied, walked round and investigated the back. Finally, having discovered the deception, he turned away in manifest disgust, and nothing that we could do or say, on that day or on any other, would induce that dog to look at that picture again. We then tried him by putting my portrait of his master also on the ground, but he simply gave it a kind of casual contemptuous sideglance and took no further notice of it. We attributed this not to any difference in the merits or demerits of the two portraits, but simply to the fact that the dog felt he had been deceived once, but was not to be so taken in again.

LOUISA STARR CANZIANI.

RECOGNITION BY ANIMALS OF PICTURES.

[*Sept.* 7, 1889.]

THIRTY years ago I was staying at Langley, near Chippenham, with a lady who was working a large screen, on which she depicted in " raised " work (as it was then called) a life-sized cat on a cushion. The host, a sportsman now dead, was much struck with the similarity to life of the cat, so he fetched his dog (alas ! like too many of the species), a cat-hater. The animal made a dead set at the (wool) cat, and but for the master's vigorous clutching him by the collar, the cushion would have been torn into atoms. I related this tale lately in Oxford, and my hearer told me that a friend in the Bevington Road had just painted a bird on a firescreen, and her cat flew at it.

My own old dog, Scaramouch (a pet of the Duke of Albany's in his undergraduate days), disliked being washed, and when I showed him a large *Graphic* picture of a

9

child scrubbing a fox-terrier in a tub, he turned his head away ruefully, and would not look at his brother in adversity.

J M. HULBERT.

DOG FRIENDSHIPS.

DOG FRIENDS.

[*Feb.* 16, 1889.]

THE following story of friendship between two dogs may, I think, interest some of your readers. Some time ago I used often to stay with a friend in Wiltshire, whose park is separated from the house by a lake which is about a hundred and fifty yards broad at the narrowest part. Being extremely fond of animals, I soon became intimate with two delightful dogs belonging to my hostess, a large collie, called Jasper, and a rough Skye terrier, Sandie. The pair were devoted friends, if possible always went out together, and, sad to relate, even poached together. One afternoon I called them, as usual, to go for a walk, and making my way to the lake, I determined to row across and wander about in the deer-park. Without thinking of my two companions, I got into the boat and pushed off. Jasper at once jumped into the water and gaily followed the boat; half way across he and I were both startled by despairing howls, and stopping to look back, we saw poor little Sandie running up and down the bank, and bitterly bewailing the cruelty of his two so-called friends in leaving him behind. Hardening my heart, I sat still in silence, and simply watched. Jasper was clearly distressed; he swam round the boat, and looking up into my face, said unmistakably with his wise brown eyes, "Why don't you go to the rescue?" Seeing, however, that I showed no signs of intelligence, he made up his mind to settle the difficulty himself, so turned and swam back to forlorn little Sandie; there was a moment's pause, I suppose for explanations, and then, to my surprise and amusement, Jasper stood still, half out and half in the water, and Sandie scrambled on to his back, his front paws resting on Jasper's neck, who swam across the lake and landed him safely in the deer-park! I need not describe the evident pride of the one, or the gratitude of the other.

Roy.

FRIENDSHIPS OF DOGS WITH OTHER ANIMALS.

A LESSON.

[Feb. 23, 1889.]

YOUR correspondent " Roy's " very interesting account of " A Canine Friendship " tempts me to send you the following about two Dandy Dinmonts in this neighbourhood.

Friends of mine in Dumfriesshire had in their house two Dandie Dinmont dogs who were inseparable friends and constant companions in all that was going on. One day one of these dogs disappeared unaccountably, and nothing was seen of it for a week. His owners were very vexed, thinking he must have got within the range of some keeper's gun or met with some other accident.

But the absentee's home-keeping companion was greatly distressed ; he moped about, and would not touch any food for several days; till, unexpectedly on my friend's part, the truant suddenly reappeared and showed himself in the house. The dog who had remained at home, when he saw the arrival of his former friend, looked steadily at him for a few seconds, and then, without further parley, went at him and gave the truant a thoroughly sound thrashing. I always explain this to myself by supposing that the home-keeping dog decided that the truant had caused him for several days needless anxiety and abstinence from food, and that the truant must learn by painful experience that such behaviour could not be lightly condoned by his inseparable companion.

J. G.

CONSCIOUS AUTOMATA.

[*July* 31, 1875.]

I HAVE lately heard a story that I hope you may think worthy of a place among your illustrations of the thoughtful intelligence of "Conscious Automata." Many years ago, a family having a house in Grosvenor Square, and a place in the country (I think in Warwickshire), owned a terrier, who, in the country, made great friends with a large Newfoundland. When they came to town they brought the terrier, and he resided in a mews where he was much annoyed by a cur who lived next door, and attacked him whenever he came out. One day the terrier disappeared, but after a little time returned, bringing with him his big friend, who gave the vulgar bully a satisfactory thrashing—not attempting to kill him. This has been told me by an old servant, who was then a young man, living in service in London, close to the owners of the dogs. He answers for the facts of the story as he heard them at the time.

F. C.

DOG AND PIGEON.

[*Sept.* 22, 1888.]

THE *Spectator* does not disdain anecdotes of dogs and their doings, and I think the following history, to which I can bear personal testimony, may be found not uninteresting to your readers. At this delightful house in Perthshire, where I am on a visit, there is a well-bred pointer, named Fop, who, when not engaged in his professional pursuits on the moor, lives chiefly in a kennel placed in a loose-box adjoining the other stables attached to the house. Nearly a year ago there were a pair of pigeons who lived in and about the stable yard. One of the birds died, and its bereaved mate at once attached itself for society and protection to the dog, and has been its constant companion ever since. On the days when the sportsmen are not seeking grouse the dog is in his kennel, and the pigeon is always his close attendant. She roosts on a rack over the manger of the stable, and in the day-time is either strutting about preening her feathers, taking her meals from the dog's biscuit and water tin, or quite as often sitting in the kennel by his side, nestling close to him. Fop, who is an amiable and rather sentimental being, takes no apparent notice of his companion, except that we observe him, in jumping into or out of his kennel while the pigeon is there, to take obvious care not to crush or disturb her in any way. The only other symptom Fop has shown of being jealous for the pigeon's comfort and convenience is that when of late

two chickens from the stable-yard wandered
into the apartment where the dog and pigeon
reside, he very promptly bit their heads off,
as if in mute intimation that one bird is
company, and two (or rather three) are none.

The story is rather one of a pigeon than
a dog, for it is quite evident that she is the
devoted friend, and that he acquiesces in
the friendship. On the days when Fop is
taken, to his infinite delight, on to the moor,
the pigeon is much concerned. She follows
him as far as she dare, taking a series of
short flights over his head, until a little wood
is reached, through which the keeper and
dogs have to take their way. At this point
her courage fails her, and she returns to the
stable, to wait hopefully for her comrade's
return.

This singular alliance is a great joy and
interest to the keepers, coachmen, and grooms
of the establishment, and as the keeper gave
me a strong hint that the story ought to be
told in print, adding that he had seen much
less noteworthy incidents of animal life promoted to such honour, I have ventured to

send it to you. I may add that the pigeon is of the kind called "Jacobin," and is white, with a black wing. Is there any precedent for such close intimacies between animals so widely separated in kind and habit?

ALFRED AINGER.

A HEN AND PUPPIES.

[*Sept.* 29, 1888.]

IN reply to Mr. Ainger's question as to there being "any precedent for such close intimacies between animals so widely separated in kind and habit" as the dog and pigeon mentioned in his interesting letter, I can mention two cases which have come under my notice this last summer at my farm in Berkshire. In one case the friendship existed between a pullet and a pig. The pullet never left the farmyard to join in the rambles of the other fowls, but kept near the pig all day, occasionally roosting on its friend's back when taking its afternoon nap.

The other case was more remarkable. A hen, with strong motherly instincts, but no family of her own, acted for several weeks as

foster-mother to eight spaniel puppies. The
real mother, a very gentle creature, soon
acquiesced in the arrangement. The hen
covered the puppies with her wings just as
though they had been chickens, and remained
with them day and night. When they began
to walk she was still their constant attendant;
when they learned to lap and eat a little she
would "call" them and break up their food.
As they grew older the poor foster-mother
had her patience sorely tried. They barked
and capered around her, leading her altogether a sad life. After the puppies deserted
her she was often seen sitting close to their
mother, the pair apparently quite understanding each other. My children were naturally
delighted to watch these strange sights, and
the hen, though not at other times very
tame, maintained perfect equanimity while
they played with the puppies around her.

F. C. MAXWELL.

A DOG AND A RABBIT.

[*Sept.* 29, 1888.]

MR. AINGER, in giving his interesting incident of strange friendships between animals, asks if there are any precedents for such incongruous intimacy as he saw between a dog and a pigeon. To most close observers of animals, such curious cases, though always noteworthy, are well known; naturalists like Buckland and many others have frequently recorded them.

With the view of adding to the lore on this matter, permit me to cite the following. Two Scotch terriers are lying before the fire. Prince is an amiable sort of dog; Jack is rather surly; both good vermin-killers and fond of hunting. I bring in a common buck rabbit, and place it beside the dogs, with the intimation they were not to touch it. Trust, and then alliance, quickly grew between it and Prince, whilst Jack shows unmistakable hatred. In a few days the two friends, with their paws absurdly clasping each other's necks, sleep happily on the rug; they play together, they chase each other up and down the stairs and all over the house at full speed, and when tired come back to the rug. Jack refusing all this sort of thing, makes the rabbit look at him with a sort of awe. Does Bunny make no mess in the house? None whatever; he goes into the garden as the dogs do, and like them, scratches at the door when he wants to return. All this he does without any instruction from us. After a while, being very fond of him, we put on the floor a pretty pink-eyed doe as a present. He stares, sniffs her all over, kills her on the spot, and goes for a romp with his dear Prince. Jack always sleeps under my bed from choice, and just before I put out the light as I lie, stands up against the bed for his last pat and "good-night." Bunny has observed all this, and quietly creeps into the room, which he refuses to leave; then likewise always asks for his "good-night," and sleeps somewhere near his great "ideal."

Another instance, published in "Loch Creran" by my friend Mr. Anderson Smith. I punished my cat for killing a chicken. The next day he is seen to carry a live chicken in his mouth and lay it down to

the hen he had previously robbed. He
and the chicken afterwards were frequently
observed leaving the orchard together, and
travelling through the courtyard and back
passages, find their way to the kitchen
fireplace, where they would sleep in good
fellowship. This chicken, I discovered, had
been stolen nearly two miles away. It is
important to remark that the cat, though
a cruel bird-killer, never touched another
chicken. Was the idea of compensation in
the cat's mind? If not that, all the circumstances are singularly coincident. And why
did the chicken prefer the cat's companionship to that of its fellows?

E. W. PHIBBS.

ANOTHER PIGEON STORY.

[Oct. 6, 1888.]

MR. AINGER's letter in the *Spectator* of
September 22nd reminds me of an almost
identical friendship that existed some years
ago at Grove House, Knutsford. A long-
haired mastiff was kept chained as a watch-
dog, and when a white fantail pigeon's mate

died, it attached itself to the mastiff, and
was continually with it in the kennel. When
the dog had its breakfast of porridge and
milk, the pigeon would eat out of the bowl
at the same time ; and when the dog had
finished, it would lie flat on its side while
the pigeon perched on its head and pecked
off the grains of oatmeal that stuck to the
long hair round its mouth. The only danger
to the pigeon seemed to be that when the dog
rushed out of the kennel suddenly to bark,
it seemed to forget the pigeon, and we used
to fear that the heavy chain might hurt it ;
but it never was hurt. This friendship
lasted many years, till one of the two, I
forget which, died.

ISABEL JAMISON.

DOG AND KITTENS.

[July 1, 1893.]

THE following story may, perhaps, interest
some of your readers :—Willie is a small,
rough-haired terrier, a truculent and aggressive character, the terror of tramps, in a
skirmish with one of whom he has lost an

eye. He rules the kitchen with a rod of
iron, the inmate there admiring and fearing
him. Next to tramps, Willie hates cats;
he has been flogged again and again for
chasing the neighbour's "Tom"; nothing can
stop him rushing at the alien cat, however.
But for his own domestic "Tabby" he has
tolerance and a certain amount of affection;
if another dog were to attack her, dire would
be the warfare. A while ago, this cat had
three kittens; two were taken by the maid
and placed in a bucket of water, and left to
their fate. Before that fate had come Willie
perceived them; he snatched them from the
bucket one by one, and carried them to his
kennel. The maid attempted to get them
away, but Willie flew at her with fury, and
then returned to lick first one and then the
other, to shove them up together, and lie
down near them, and in every way to give
the poor half-dead things a chance. This
went on for some time; but when at last
there was no sign of breath, and he saw that
they were hopelessly dead, he marched out of
the kennel, shook himself, and indicated to

the maid that she might now proceed to bury them, that they were past intelligent treatment. He treats the remaining and living kitten with the indifference of the scientific for the normal.

L. H.

A CANINE NURSE.

[*May* 18, 1895.]

BEING a frequent reader of anecdotes of the sagacity of animals in your paper, I think you may consider the following trait of character in a dog worthy of notice. Jack, a rough-haired fox-terrier of quiet disposition, but a good ratter, and an inveterate enemy to strange or neighbouring cats, of whom, to my sorrow, he has slain at least one, became without effort the attached friend of a minute kitten introduced into the house last November. This friendship has been continued without intermission, and is reciprocated by the now full-grown cat. She, unfortunately, got caught in a rabbit-trap not long ago, but escaped with no further injury than a lacerated paw, which for some time caused her much pain and annoyance. Every morning Jack was to be seen tenderly licking the paw of the interesting invalid, to which kind nursing no doubt her rapid recovery may be attributed; and though she is now more than convalescent and able to enjoy her usual game of play, he still greets her each morning with a gentle inquiring lick on the injured paw, just to see if it is all right, before proceeding to roll her over in their accustomed gambols. This seems to me a marked instance of individual affection overcoming race-antipathy.

BLANCHE ROCHFORT.

A CURIOUS FRIENDSHIP.

[*Feb.* 6, 1875.]

I HAVE two dogs, two cats, and a kitten. Many years of experience have shown me, in the teeth of all proverbs, that cats and dogs, members of the same household, live together quite as amicably as human beings.

Only, like human beings, they have their dislikes and preferences for each other. At the present time, my dog Snow is on terms of hearty friendship with my grey cat Kitty, but of polite indifference with my black cat Toppy.

Toppy, for some years back, has been subject to fits, owing, it is considered, to the lodgment of some small shot near her spine, whilst out trespassing (or poaching).

Yesterday Snow rushed into the kitchen with face so anxious and piteous that my servants both exclaimed that something must have happened ; gave signs, as he can do, that somebody was to go with him, and was followed into the drawing-room, where Toppy, left alone, had fallen under the grate in a fit, and was writhing amid the ashes and embers. She was rescued, and beyond a little singeing, does not seem much the worse.

To reach the kitchen, Snow must have pushed open a red-baize door, which he has never been known to open before, and before which he will stay barking for ten minutes at a time to be let through.

If any biped, supposing himself to be endowed with reason, humanity, and articulate speech, tells me that Snow is a conscious automaton, can I give him any other answer than, "You're another"?

J. M. L.

AN ACT OF CANINE FRIENDSHIP.

[*Nov.* 6, 1880.]

I HAVE read from time to time in the pages of the *Spectator* instances of canine sagacity furnished by your correspondents, which have, no doubt, interested many others besides myself. The following incident occurred last Saturday, in my walk from the beach, which, perhaps, may amuse your readers, as it did me.

My curiosity was excited by seeing a young retriever on his hind legs licking very ardently the face of a nice-looking donkey, who was tethered on the bank. After licking his face all over for a long time, he began to frisk around him, evidently anxious to have a trot together; but, finding that his friend was tied by a rope, he deliberately began to gnaw it, and in a very short time succeeded in setting him free! The owner

of the donkey, who happened to be at work
close by, then interfered, and put a stop to
their little game, or otherwise Master Neddy
would, no doubt, have been seduced to join
in a scamper. From the warmth of the
dog's salutes, I imagine that he and the
donkey were old friends.

S. RICHARDS.

DOG AND CANARY.

[*Nov.* 20, 1880.]

I WAS much interested in the account of the
friendship that existed between the young
retriever and the donkey whom he released
by gnawing the rope. The little incident
I send of another retriever may also interest
your readers. A friend of mine had a pet
canary, while her brother was the owner
of a retriever that was also much petted.
One day the canary escaped from the house,
and was seen flying about the grounds for a
few days, and when it perched was generally
on high elm-trees. At last it vanished from
view, and this dear little pet was mourned
for as lost or dead. But after the interval

of another day or so, the retriever came in with the canary in his mouth, carrying it most delicately, and went up to the owner of the bird, delivering it into her hands without even the feathers being injured. Surely nothing could illustrate more beautifully faithful love and gentleness in a dog than this.

E. TILL.

CAT-AND-DOG LOVE.

[*April* 13, 1878.]

WOULD you allow me, as a cat fancier of nearly thirty years' standing, to corroborate, by a personal experience, Mr. Balfour's testimony in your last issue to the possibility of a genuine attachment between a cat and a dog? A few weeks ago, I called upon a bachelor friend who has two pets, a handsome black female cat, of the name of Kate, and a bright little terrier, responding to the call of David. My friend assured me that they lived on the most affectionate terms. They were certainly not demonstrative, but they were importations from Scotland, and refrained from " spooning " before folk. The character of the attachment was soon tested. Another acquaintance entered the room, accompanied by a terrier of about the same size as David, although not of the same variety. This dog made at once for the cat, then resting in front of the fire. She backed against the wall, and prepared for a fight, in which, if I may judge from her size, she would have been victorious. But she was saved the trouble of using her claws. Before she could utter a feline equivalent for "Jack Robinson," before the door could be closed, David rushed at the intruder, and literally ran him out of the room and down two flights of stairs, with a rapidity worthy of a member of the Irish Constabulary. By the time he returned, his Dulcinea had arranged herself for another nap, but she opened one eye as her companion took his place by his side, and—

" Betwixt her darkness and his brightness,
    There passed a mutual glance of great politeness."

I witnessed a similar scene some years ago in a country inn in the north of Scotland.

On that occasion, one dog defended against another a favourite cat and a favourite hen.

Speaking of cats, can any one say what has become of the late Pope's black cat, Morello? Did he die before his master, or has some one adopted him? Châteaubriand, as everybody knows, adopted Micetto, the grey favourite of Leo XII.

WILLIAM WALLACE.

THE DOG THAT BURIED THE FROGS.

[*Feb.* 2, 1895.]

KNOWING your love of animals, and the interest so often shown in your columns in their ways, I venture to send you the following story I have lately heard from an eye-witness, and to ask whether you or any of your readers can throw any light upon the dog's probable object. The dog in question was a Scotch terrier. He was one day observed to appear from a corner of the garden carrying in his mouth, very gently and tenderly, a live frog. He proceeded to lay the frog down upon a flower-bed, and at once began to dig a hole in the earth, keeping one eye upon the frog to see that it did not escape. If it went more than a few feet from him, he fetched it back, and then continued his work. Having dug the hole a certain depth, he then laid the frog, still alive, at the bottom of it, and promptly scratched the loose earth back into the hole, and friend froggy was buried alive! The dog then went off to the corner of the garden, and returned with another frog, which he treated in the same way. This occurred on more than one occasion; in fact, as often as he could find frogs he occupied himself in burying them alive. Now dogs generally have some reason for what they do. What can have been a dog's reason for burying frogs alive? It does not appear that he ever dug them up again to provide himself with a meal. If, sir, you or any of your readers can throw any light on this curious, and for the frogs most uncomfortable, behaviour of my friend's Scotch terrier, I should be very much obliged.

R. ACLAND-TROYTE.

AN EXPLANATION.

[*Feb.* 9, 1895.]

I THINK I can explain the puzzle of the Scotch terrier and his interment of the frogs, for the satisfaction of your correspondent. A friend of mine had once a retriever who was stung by a bee, and ever afterwards, when the dog found a bee near the ground, she stamped on it, and then scraped earth over it and buried it effectually—presumably to put an end to the danger of further stings. In like manner, another dog having bitten a toad, showed every sign of having found the mouthful to the last degree unpleasant. Probably Mr. Acland-Troyte's dog had, in the same way, bitten a toad, and conceived henceforth that he rendered public service by putting every toad-like creature he saw carefully and gingerly "out of harm's way," underground.

A great number of the buryings and other odd tricks of dogs must, however, I am sure, be considered as Atavism, and traced to the instincts bequeathed by their remote pro-

genitors when yet "wild in the woods the
noble *beastie* ran." Such, I believe, is
generally admitted to be the explanation of
the universal habit of every dog before lying
down to turn round two or three times and
scratch its intending bed—even when that
bed is of the softest woollen or silk—
apparently to ascertain that no snakes or
thorns lurk in its sleeping-place.

A dog which I once possessed exhibited
such reversion to ancestral habits in a noteworthy way. She was a beautiful white
Pomeranian; and when a litter of puppies
was impending, on one occasion she scratched
an enormous hole in our back-garden in
South Kensington, where her leisure hours
were passed—a hole like the burrow of a
fox. It was not in the least of the character
of the ordinary circular punch-bowl so often
scooped out by idle or impatient dogs,
but a long, deep channel running at a sharp
angle a considerable way underground.
Obviously, it was Yama's conviction that
it was her maternal duty to provide shelter
for her expected offspring, precisely as a

fox or rabbit must feel it, and as we may
suppose her own ancestresses did on the
shores of the Baltic some thousand genera-
tions ago. When the puppies were born,
Yama and the survivor were established by
me in a most comfortable kennel in the
same garden, with a day nursery and a
night nursery (covered and open) for the
comfort and safety of the puppy. But one
fine morning, when the little creature had
begun to crawl over the inclosure of its small
domain, I happened to go into the garden
while Yama was absent in the house, and
discovered that my little friend was missing.
The puppy had disappeared altogether ; and
at the same time I noticed that the flower-
bed in which Yama had made her excavation
had been nicely smoothed over by the
gardener, who was putting the place in
order. A suspicion instantly seized me, and
I exclaimed, "You have buried my puppy ! "
I ran to the spot where the hole had been
made, and, having swept aside the gardener's
spadeful of soil, found the deeper part of the
hole, running slanting underground, still

open. I knelt down and thrust in my arm to its fullest stretch, and then, at the very end of the hole, my fingers encountered a little soft, warm, fluffy ball. The puppy came out quite happy and uninjured, freshly awakened from sleep, having shown that his instinct recognised the suitability of holes in the ground for the accommodation of puppies; just as the hereditary instinct of his mother had led her to prepare one for him, even in a South Kensington garden !

FRANCES POWER COBBE.

A DOG AND HIS DINNER.

[*Feb.* 16, 1895.]

I KNEW a dog in Ireland—a large retriever—who had been taught always to bring his own tin dish in his mouth, to be filled at the late dinner. For some reason his master wished to make a change, and to feed him twice a day instead of once, to which he had always been accustomed. The dog resented this, and when told to bring his dish, refused, and it could nowhere be found ; on which his master spoke angrily to him, and ordered him to bring the dish at once. With drooping tail and sheepish expression he went down the length of the garden, and began scratching up the soil where he had buried the bowl deep down, to avoid having to bring it at an hour of which he did not approve.

A LOVER OF DOGS.

DOGS AND LOOKING-GLASSES.

[*June* 23, 1894.]

You are fond of odd actions of dogs, so perhaps the following may be acceptable. I have two fox-terriers—young dogs—Grip and Vic. In the morning, at early tea in our bedroom, Vic gets angry with Grip's reflection in the long glass of the wardrobe, barks at him furiously as he moves about, and scratches at the glass, quite regardless of her own face between her and his reflection. And when he assaults her from behind, to make her play with his real self, she turns round and snaps at him viciously, and then returns to her attack on his reflection. He jumps upon the window-sill, and fancies he sees a squirrel in the garden, and dashes past her to the door ; she follows the motion of the reflection till she is past the edge of the glass, and loses it, when she dashes back to the glass again. This has occurred several days in the last week, and seems to me almost absurd. The dogs are just about a year old, and so beyond puppy folly, though very lively and playful still.

A. M. B.

THE SENSE OF HUMOUR AND CUNNING IN DOGS.

DOGS' SENSE OF HUMOUR.

THE POWER OF IMITATION IN DOGS.

[*Oct.* 22, 1882.]

THE following anecdote may interest those of your readers who are accustomed to observe the characteristic actions of dogs. I can vouch for its accuracy, as I was an amused eye-witness, and several members of my family were also present, and have often told the story.

A friend of ours and his wife were spending a musical evening with us, and an old, black, English terrier, who belonged to the house, had been in the drawing-room, which was upstairs. The dog had been kindly noticed by our friend, who was partially lame from paralysis. On leaving the drawing-room the dog followed him to the top of the staircase (we, with his wife, were waiting below in the hall), and with cocked tail and

ears stood gravely watching his slow, limping
descent. When the invalid was nearly at
the foot of the stairs the dog began to follow,
limping on three legs (he was quite sound),
in humorous imitation of our poor, afflicted
friend, and this assumed lameness was gravely
kept up till he arrived on the mat. It was
impossible to repress a smile, though our
politeness was at stake, and the unconscious-
ness of our friend added to the difficulty.

A. R.

SENSE OF HUMOUR IN DOGS.

[*July* 28, 1888.]

A RECENT anecdote from one of your correspondents about a dog and a hen brought to
my mind an incident, related to me by an
eye-witness, of a dog who had a constant
feud with the fowls, which were prone to
pilfer from the basin containing his dinner.
On one occasion he was lying in front of his
kennel, quietly watching a hen as she made
stealthy and tentative approaches to his
basin, which at length she reached and

looked into, finding it perfectly empty.   The
dog wagged his tail.

J. R.

A DOG'S SENSE OF HUMOUR.

[*March* 9, 1895.]

DOES the following dog-story show a sense
of humour ?   A retriever was in the habit of
leaving his bed in the kitchen when he heard
his master descending the stairs in the morning.   On one occasion a new kitchen-maid
turned him out of his bed at a much earlier
hour than usual.   He looked angrily at her,
but walked out quietly.   Time passed, and
he was nowhere to be found.   At last, in
going to her bedroom, the kitchen-maid
found him coiled up in her own bed.

B. B.

CUNNING DOGS.

A DOG AND A WHIP.

[*May* 18, 1889.]

You have lately published several dog stories. Allow me to send you another for publication should you think it worthy. It was told me to-day by a lady whom I cross-examined to get full details:—" Some twenty years back we had a poodle—white, with one black ear. After the manner of his race, he was never quite happy unless he carried something in his mouth. He was intelligent and teachable to the last degree. The great defect in his character was the impossibility of distinguishing *meum* from *tuum*. Anything he could get hold of he seemed to think, according to his dogged ethics, to be fairly his own. On one occasion he entered the room of one of the maidservants and stole her loaf of bread, carefully shutting the door after him with his feet—the latter part being a feat I had taught him. The woman —Irish—was scared, and thought that the dog was the devil incarnate. The necessity of discipline on the one hand, and of occupation on the other, induced me one day to enter a saddler's shop, situated in a straight street about half a mile from our house, and buy a whip. Shortly after my return home he committed some act of petty larceny, so I gave him a beating with the whip he had carried home. Going for a walk next day the dog, as usual, accompanied me, and was entrusted with the whip to carry. Directly we got outside the door he started off at his best pace straight down the street, paying no attention whatever to my repeated calls. He entered the saddler's shop and deposited the whip on the floor. When I arrived the saddler showed me the whip lying exactly where the dog had deposited it."

HENRY H. MAXWELL.

A RUSÉ DOG.

[*March* 21, 1885.]

A STORY which came to my knowledge a few months ago may be of interest in connection

with the *Spectator's* series of anecdotes illustrating the intelligence of animals.

One summer afternoon a group of children
were playing at the end of a pier which
projects into Lake Ontario, near Kingston,
New York, U.S.A.   The proverbial careless
child of the party made the proverbial back-
ward step off from the pier into the water.
None of his companions could save him, and
their cries had brought no one from the shore,
when, just as he was sinking for the third
time, a superb Newfoundland dog rushed
down the pier into the water and pulled the
boy out.   Those of the children who did not
accompany the boy home took the dog to a
confectioner's on the shore, and fed him with
as great a variety of cakes and other sweets
as he would eat.   So far the story is, of
course, only typical of scores of well-known
cases.   The individuality of this case is left
for the sequel.

The next afternoon the same group of
children were playing at the same place,
when the canine hero of the day before came
trotting down to them with the most friendly

wags and nods. There being no occasion this time for supplying him with delicacies, the children only stroked and patted him. The dog, however, had not come out of pure sociability. A child in the water and cakes and candy stood to him in the close and obvious relation of cause and effect, and if this relation was not clear to the children he resolved to impress it upon them. Watching his chance, he crept up behind the child who was standing nearest to the edge of the pier, gave a sudden push, which sent him into the water, then sprang in after him, and gravely brought him to shore.

To those of us who have had a high respect for the disinterestedness of dogs, this story may give a melancholy proof that the development of the intelligence, at the expense of the moral nature, is by no means exclusively human.

Clara French.

DOG DECEIVERS.

[*Feb.* 9, 1895.]

Your fondness for dogs induces me to send

you the following anecdote, which shows
their power of acting a part for purposes of
their own. Some years ago a fox-terrier of
mine was condemned by a veterinary surgeon to consume a certain amount of flour
of sulphur every day. He was at all times
a fanciful and dainty feeder, and every con-
ceivable ingenuity on my part was exhausted
in the vain endeavour to disguise the daily
portion and to give it a more tempting appearance. Each new device was invariably
detected. However hungry he might be
he turned from the proffered morsel in disgust, and it ended almost invariably in my
having to put it down his throat. One
morning, after keeping him for many hours
without food, and having neatly wrapped the
powder in a most appetising piece of raw
meat, I offered it him in the vain hope that
hunger might prevail over prejudice. But
no. With averted head and downcast look
he steadily and determinedly declined to par-
take of it. I encouraged him in vain. Deep
dejection on his part; despair, but persistence,
on mine. All of a sudden his whole manner

changed. He assumed a brisk and cheerful demeanour, joyfully accepted the hitherto rejected offering, and running merrily through the open door, disappeared swiftly a few yards off round the corner of the building. Inside the room I ran as quickly to a window, whence I could view his proceedings, and there watched him while he deposited the hated morsel on the ground, dug a hole in the flower-bed, and buried it. His jaunty, triumphant air as he returned I shall never forget.

F. E. WYNNE.

USEFUL DOGS.

GUARDIAN DOGS.

[*July* 15, 1892.]

HAVING read for years your interesting letters and articles on animals in the *Spectator*, I feel sure you will like to have a thoroughly authentic account of a dog in this neighbourhood. I am allowed to give the name of the owner, who is living at Lyme Regis, where I was staying last week. The two incidents happened within a few weeks of each other.

Mrs. and Miss Coode were alone in their house (except the servants); and one night Miss Coode was awakened by hearing two knocks at her door and a slight whine. It was between three and four o'clock in the morning. She rose and opened the door to find the dog there, and at the same time noticed and heard a stream of water running down the stairs. She went up the staircase to its source, and aroused the servants to attend to it. As soon as the dog saw that the matter was being remedied, he quietly went back to the mat in the hall and went to sleep again. The dog is a large one, a cross between a retriever and a greyhound — a very beautiful creature, resembling a poacher's lurcher.

The second incident occurred only last week, when Miss Coode was again aroused. This time by a loud crash, as if a picture had fallen. Almost immediately the dog bounded upstairs, threw himself against the door, which happened to be ajar, burst into the room, panting and eyes glistening,—this, at least, Miss Coode saw as soon as she struck a light, for it was between twelve and one o'clock. She went out on to the staircase and downstairs to look at the pictures in the drawing-room. The dog would not follow. The cook, coming down from her room, called him a coward not to go with his mistress, but Sheppard did not move. Miss Coode found all safe below, and returned upstairs, and the dog went with her to the top floor, where the ceiling of a small room had fallen in. He then retired to his mat, having done his duty. He also showed his sagacity in going to the daughter's room—the one most capable of seeing to matters. Hoping, as a dog-lover, that this may interest all such, and help to prove that dogs think and reason more than some human beings—also to show that we often inferior beings have no right to presuppose that the superior animals have no souls.

K. Clarke.

A TRUE WATCH-DOG.

[*Aug.* 5, 1893.]

The "dog" letter in the *Spectator* of July 15th is wonderfully like my experience, some years ago, with my little red Blenheim, Frisk. She always slept in a basket, close to the hall door. One night she dashed up the stairs, loudly barking, ran first to my eldest sister's room, then through a swing-door to another sister's room, barking outside each door, then upstairs again to my room at the top of the house, where she remained barking

till I got up and opened it, when she ran in,
still barking, and waited till I was ready to
go down with her.   She scampered on before
me, I following close, and when we both
reached the hall she dashed still barking to
the door, to show me whence her alarm had
arisen.   It was the policeman turning the
handle of the door from the outside to see if
it was properly closed!   One night, a long
time after the first adventure, I was wakened
by a quiet scratch at the door of my room.
No barking this time ; but, tiresome as it
was to be disturbed on a cold night, I got up
and opened the door, and was conscious in
the darkness that Frisk was standing there.
"Come in, Frisk," said I.   But no movement ; Frisk stood waiting.   "Come in,
Frisk," I repeated, somewhat sharply.   No
movement, no bark!   Then, being sure that
something must be wrong, I lighted a candle,
and there stood Frisk outside the door,
never offering to come in.   She trotted
quietly down before me, not speaking a
word.   When we were both through the
swing-door, and at the head of the stairs,

I saw that the inner door to the hall was open, and also that of the morning-room, from which shone a bright light. My heart went pit-a-pat for a moment ; then seeing Frisk run quietly down the stairs, I followed her, when she calmly jumped into her basket again, and I, venturing into the morning-room, found that my brother-in-law had left the lamp burning by mistake—a proceeding which Frisk plainly knew was wrong, and had therefore come upstairs to inform me, but had not thought it necessary to disturb the rest of the household this time! She had come straight up to my room without disturbing any one else, to tell me of the irregularity of a light burning when every one was in bed, and that being done, jumped into bed again, conscious of having performed her duty.

Georgina A. Marsh-Caldwell.

[Aug. 12, 1893.]

I can give an instance as convincing as that of Miss Marsh-Caldwell of the way in which a true watch-dog will measure the extent of his duties. I lived for many years opposite a wood, in which the game at first was preserved. I had a dog named Prin, who had begun by being a gardener's dog, but having caught the distemper and been unskilfully treated by his master he remained nearly blind, and was left on my hands by the man when he quitted my service. The dog was a great coward, but good-tempered and affectionate, and the partial loss of sight seemed to have developed greatly the senses both of hearing and smell, so that he was recognised as a capital watch-dog. He was promoted to the kitchen, and would have been promoted to the drawing-room but for the obstreperousness of his affection, which seemed to know no bounds if he was admitted even into the hall. I slept at that time in a room over the kitchen, fronting the road. One night I was awakened by Prin growling, and, after a time, giving a snappish bark underneath me. I got out of bed and throwing up the sash, listened at the window, where, after a time, I heard slight noises, which convinced me that some one or more persons were hiding in the shrubbery between the house and the road, whom I supposed to be burglars. I called out, "Who's there?" without, of course, eliciting any answer, and, after a time, I heard the click of the further gate (there being two, one opposite my house, the other opposite its semi-detached neighbour, and out of my sight), after which all was quiet. But I had noticed that from the moment of my getting out of bed Prin had not uttered a sound. The same thing happened seven or eight times, and always in the same way, Prin growling or barking till he heard me get out of bed, and then holding his tongue, as feeling that he had fulfilled his duty in warning his master, and that all responsibility now devolved upon me. The secret of the matter I discovered to be that poachers, with no burglarious intentions towards me, used the shrubbery as a hiding-place before getting over the opposite paling into the wood.

One other instance of Prin's sagacity I will also mention. I had a black cat, with white breast, named Toffy, between whom and

Prin there was peace, though not affection.
There was also another black cat, with white
breast, that prowled about, an outlaw cat,
who made free with my chickens when he
could! It was a bitter winter, and the snow
had lain already for days on the ground.
I was walking one Sunday morning in my
garden, Prin being out with me. He quitted
me to go under a laurel-hedge bounding a
shrubbery, and presently began barking
loudly. I went towards him, and saw a
white-breasted cat sitting stretched under
the laurels, with front paws doubled under
him, which I took to be Toffy asleep. I
scolded Prin for disturbing Toffy, and he
stopped barking, but remained on the spot
whilst I continued my walk. Presently—
say two or three minutes after—I heard him
barking still more loudly than before, and so
persistently that I returned to the spot.
Noticing that the cat had never moved
through all the noise, I crept up under the
bushes, and found that it was not Toffy
asleep, but the outlaw cat, dead—evidently
of cold. Thus my poor purblind watch-dog

had—(1) barked to draw my attention to what appeared to him an unusual phenomenon ; (2), held his tongue in deference to my (supposed) superior wisdom, when I told him he was making a mistake ; (3), not being, however, satisfied in his mind, remained to investigate till he was convinced he had not been mistaken ; (4), called my attention to the facts still more instantly till I was satisfied of them for myself. Could *homo sapiens* have done more ?

J. M. L.

[*Aug.* 12, 1893.]

I AM reminded by the anecdote related in the *Spectator* of July 15th, "A Canine Guardian," of the sagacity of a favourite Scotch terrier which was displayed some years ago. I was dressing one morning, and my bedroom-door was ajar. Standing at my dressing-table, I was surprised to see Fan come up to me, frisking about, and looking eagerly into my face, whether from pleasure or not I could not tell. I spoke to and stroked her, but she was in no way soothed, and she ran out

of the room evidently much excited.   In she
came again, more earnestly trying to tell me
what she wanted, rushing up to me and again
to the door, plainly begging me to follow her,
which I did, into the next room, where breakfast was laid.   I at once saw what she had
easily felt was out of order—the kettle was
boiling over, and the water pouring from the
spout had drenched the hearth.   Hence her
discomfort, and her effort to tell me of the
disaster.   Having brought me on the scene,
she seemed perfectly content.

C. A. T.

[*Aug.* 12, 1893.]

NOT long ago I was passing a barn-yard in
this place, and stood to look over the gate at
a pretty half-grown lamb standing alone outside the barn.   But the sight of me so
enraged a fierce, shaggy grey dog tied up
to his kennel between the lamb and me,
that he barked himself nearly into fits,
showing all his teeth, and straining so
furiously at his chain as to make me quite
nervous lest it should give way.   In the

meantime, I struck such terror into the
heart of the lamb that it fled across the
yard to place itself under the protection of
the dog, and stood close by his side, whilst
he barked and danced with fury. As I drew
a little nearer, the lamb backed right into the
kennel, and when, after I had made a circuit
in order to watch the further movements of
this strange pair of friends from behind a
tree, I saw their two faces cautiously looking
out together, cheek-by-jowl, whilst the dog's
anger was being reduced to subsiding splut-
ters of resentment. He was not a collie, but
a very large sort of poodle.

C. S.

COLLIES AT WORK.

[March 25, 1893.]

AT six o'clock this morning, I saw a mountain-shepherd stand at a gate on the hill-top. Seven sheep were on the outside of the gate—six of the shepherd's flock, the other a strayer. The man wanted his own sheep in; so, before opening the gate, he quietly said : "'Rob,' catch the strayer." In an instant "Rob" pinned the sheep, holding him, strong and wild as he was, as though he were in a vice; and then, by another word, "Gled" was told to bring the others in through the gate now opened for them. Although "Gled" brought his six wild sheep right over "Rob" and his strayer, the sheep was held securely till the gate was closed, and the order given to "let it gang."

WILLIAM FOTHERGILL.

A COLLIE AT WORK.

[*Aug.* 11, 1894.]

WE stood at the bottom of a deep valley
with the hills rising abruptly on either side,
when Robert Scott said: "Yonder is the
sheep I led away from Llangynider, all those
weary miles yesterday. I saw it as I came
over the hill-top down to the house this
morning. If you wish, "Kate" shall bring
it down to my feet here for you to see it."
"What?—bring that single sheep! How
will she know the one you want, and how
can she get it away from the flock by
itself? I will not believe that possible till
I see it done, at all events."

He spoke a low word or two to the
collie by his side, and away went "Kate"
right up over rock and bracken, till we
could see the flock far away upon the height
above give a very rapid turn, and in a few
minutes afterwards, down rushed a strong
mountain wether with the wily "Kate"
working to the right and left about thirty
yards behind it. "Come away, back 'ahint

me," cried Scott ; and "Kate," at once leaving
the sheep, appeared positively to fly far
out, and coming round behind us, stopped
the wether in his headlong course, bringing
him to a stand literally at the shepherd's
feet. "Robert," I said, "when (as you
intend) you sail next month for New
Zealand, you will not take 'Kate' with
you, but leave her here for seven sovereigns." "Nae, nae, sir," was the reply,
"seventy sovereigns would nae buy her."

W. Fothergill.

MISCELLANEOUS.

13

A SUNDAY DOG.

[*Feb.* 17, 1877.]

A CORRESPONDENT favoured your readers last week (see page 53) with an interesting anecdote of a dog's intelligence in reference to the use of money. Permit me to relate an instance of a dog's intelligence in reference to the day of the week. Some three-and-twenty years ago, in the infancy of the Canterbury Province, New Zealand, there lived in the same neighbourhood as myself two young men, in the rough but independent mode of life then prevalent in the colony, somewhat oblivious of old institutions. These men possessed a dog each, affectionate companions of their solitude. It was the custom of this primitive establishment to utilise the Sabbath by a ramble, in quest of wild ducks and wild pigs, about the swamps and creeks of the district. It was observed that long before any preparations were made for starting, the dogs always seemed to be more or less excited. This was remarkable enough, but not so much as what followed. One of these men after a while left his friend, and taking his dog with him, went to live with a clergyman about four miles off. Here ducks and pigs had to be given up on Sundays for the church-service. It was soon noticed that this dog used to vanish betimes on Sundays, and did not turn up again until late. Upon inquiring, it was found that the dog had visited its old abode, where on that day of the week sport was not forbidden. The owner tried the plan of chaining up the animal on Saturday evenings, but it soon became very cunning, and would get away whenever it had the chance. On one occasion it was temporarily fastened to a fence-rail about mid-day on a Saturday. By repeated jerks it loosened the rail from the mortice-holes, and dragged it away. Upon search being made, this resolute but unfortunate dog was found drowned, still fast to the chain and rail, in a stream about two miles away in the direction of its old haunts. The gentleman who owned the other dog is in England now, and went over the details of the facts herein stated with me quite recently.

ALFRED DURELL.

A COW'S JEALOUSY OF A DOG.

[April 30, 1892.]

As a subscriber to and constant reader of the *Spectator*, I have derived much pleasure from the anecdotes of animal instinct, sagacity, and emotion, which from time to time have appeared in your columns. Perhaps you may like to publish the following instance of jealousy in a cow; it is, at any rate, a story at first-hand, as I myself was an actor in the affair.

A few years ago, I had a quiet milch-cow, Rose, who certainly was fond of Thomas, the man who milked her regularly, and she also showed an aversion to dogs even greater than is usual in her species. One night, for what reason I now forget, I had tied up a young collie dog in the little cowshed where she was accustomed to be milked. The following morning, I had just begun to dress, when I heard the puppy barking in the cowshed. "Oh!" thought I, "I forgot to tell Thomas about the puppy, and now the cow will get in first and gore it." The next minute I heard a roar of unmistakable fear and anguish— a human roar. I dashed down to the spot, and at the same moment arrived my son, pitchfork in hand. There lay Thomas on his face in a dry gutter by the side of the road to the cowhouse, and the cow butting angrily at him. We drove off the cow, and poor Thomas scuffled across the road, slipped through a wire fence, stood up and drew breath. "Why, Thomas," said I, "what's the matter with Rose?" "Well, sir," said Thomas, "I heard the pup bark and untied him, and I was just coming out of the cowhouse, with the pup in my arms, when 'Rose' came round the corner. As soon as she see'd the pup in my arms, she rushed at me without more ado, knocked me down, and would have killed me if you hadn't come up." Thomas had indeed had a narrow escape; his trousers were ripped up from end to end, and red marks all along his legs showed where Rose's horns had grazed along them. "Well," said I, "you'd better not milk her this morning, since she' in such a fury." "Oh! I'll milk her right enough, sir, by and by ; just give her a little time to settle down like. It's only jealousy of that 'ere pup, sir. She couldn't abide seeing me a-fondling of it." "Well, as you like," said I ; only take care, and mind what you're about." "All right, sir!"

In about twenty minutes, Thomas called me down to see the milk. The cow had stood quiet enough to be milked. But the milk was deeply tinged with blood, and in half an hour a copious red precipitate had settled to the bottom of the pail. Till then I had doubted the jealousy theory. After that I believed.

C. HUNTER BROWN.

AN AUSTRALIAN DOG-STORY.

[May 11, 1895.]

SEEING the great interest which many of your readers take in the study of canine character and intelligence, I think perhaps the following incident is worth recording. Whilst walking with a lady friend along Studley Park Road, Kew (a residential suburb of Melbourne), on a very quiet afternoon some time ago, we were surprised by a large St. Bernard dog, which came up to us and deliberately pawed my leg several times. Our perplexity at his extraordinary behaviour was perhaps not unmixed with a little misgiving, for he was an animal of formidable size and strength ; but as he gave evident signs of satisfaction at our noticing him, and proceeded to trot on in front—at intervals looking round to make sure we were following—we became interested. When we had followed him about forty yards, he stopped before a door in a high garden wall, and, looking round anxiously to see that we were noticing,

reached up his paw in the direction of the
latch. On stretching forth my hand to
unfasten the door, his extreme pleasure was
exhibited in a most unmistakable manner;
but when he saw me try in vain to open
it, he became quiet, and looked at me with
an expression so manifestly anxious that
I could no more have left the poor animal
thus than I could have left a helpless little
child in a similar position. With eager
attention and expectancy he listened while
I knocked, and when at last some one was
heard coming down the garden path, he
bounded about with every sign of unlimited
joy.

Now here was one of the so-called
"brutes," which, failing to get in at a
certain door, cast about for a way out of
the difficulty, and seeing us some distance
down the road (we were the only persons
in sight at the time), he had come to us,
attracted our attention, taken us to the door,
and told us he wanted it opened. We both
agreed that the animal had all through
shown a play of emotion and intelligence

comparable to that of a human being ; and, indeed, we felt so much akin to the noble creature that we have both, since then, been very loath to class dogs as "inferior animals."

GEORGE EASTGATE.

TWO ANECDOTES OF DOGS.

[*Feb.* 2, 1895.]

HAVING derived much pleasure from reading
the frequent natural history notes which
from time to time appear in the *Spectator*, I
venture to send you two instances of what
seems to me the working of the canine
mind under quite different circumstances.
The first refers to an incident which happened a great many years ago. It was this.
One day, when a lad, I was walking with my
father accompanied by a strong, smooth-
haired retriever called Turk. We were
joined by the bailiff of the farm, and in the
course of our walk Turk suddenly discovered
the presence of a rabbit concealed in what in
Scotland is called a "dry-stane dyke." After
a little trouble in removing some stones, poor
bunny was caught and slaughtered, being
handed to the bailiff, who put it in his coat
pocket. Shortly afterwards we separated,
the bailiff going to his home in one direction, and we to ours in an opposite one.
Before we reached home we noticed that

Turk was no longer with us, at which we were rather surprised, as he was a very faithful follower. Some time after we got home, perhaps an hour, I chanced to see a strange object on the public road which puzzled me as to what it was. It raised a cloud of dust as it came along, which partly obscured the vision. What was my surprise when I found it was Turk dragging a man's shooting-jacket, which proved to be the bailiff's, with the rabbit still in the pocket. We afterwards learnt that the dog, to the surprise of the bailiff, quietly followed him home, and lay down near him. Presently the man took off his coat, and laid it on a chair. Instantly Turk pounced upon it, and dashed to the door with it in his mouth. He was pursued, but in vain, and succeeded in dragging the coat from the one house to the other, a distance of one mile and three-fourths. It was evident the dog had a strong sense of the rights of property. He believed the rabbit belonged to his master, so he set himself to recover what he thought stolen goods.

The other anecdote refers to quite a recent date, and the only interest it has, is that it shows how perfectly a dog can exhibit facial expression, and also read at a glance the slightest indications of feeling in the human face. I had a well-broken Irish setter, which was perfectly free of hare or rabbit as to chasing, but he was a sad rascal for all that. I also had at the time a rough Scotch terrier, and the two dogs were great chums. The moment they got the chance they were off together on a rabbit-hunt. Like idiots, they would spend hours in vainly trying to dig rabbits out of their burrows. One day as I was returning home I met the pair in the avenue. They were the very picture of happiness. At first they did not see me, and came joyously on at a trot. The instant they observed me they came to a full stop, some forty yards off. The setter gently wagged his tail, and looked at me with an expression of anxious inquiry. Taking heart, he slowly advanced to within about thirty yards, and then came the varying play of feature which so interested me. He was in

great doubt as to whether I had guessed
what tricks he had been up to ; but as I
made no sign, he was gradually looking
more comfortable and gaining confidence.
Suddenly I noticed a patch of mud above
his nose, and I must have unconsciously
shown him I had made a discovery of some
kind, for that instant he turned tail and
bolted home at the utmost speed of which
he was capable. Without uttering a single
word, or making a single gesture, the dog
and man understood each other perfectly.
It was the language of faces.

R. Scot Skirving.

A DOG OBEYING A SUMMONS

[*Jan.* 18, 1890.]

THE enclosed may interest you. I received it this morning. I have no doubt Dr. Barford, of Wokingham, would verify it, but I have not the pleasure of his acquaintance. The following is the story :—

" Dr. Barford's dog at Wokingham was put into a muzzle ; he objected to it, took it off, and hid it somewhere, no one knows where. Policeman saw him ; summoned Dr. B. ; case was to come off one Saturday. The children told dog how wicked he'd been : Dr. B. would have to appear at the Court, and he too, as it was his doing ; *he'd* lost the muzzle. Case was postponed (I think policeman witness had influenza). Dr. B. was told of postponement by letter ; forgot to tell children or dog. At Saturday's Bench, Magistrates much astonished by the dog appearing in Court and sitting solemnly opposite them."

ALYS M. WOOD.

A PUG'S INTELLIGENCE.

[*Feb.* 1, 1890.]

SEVERAL newspaper cuttings have been sent
to me with the story of my dog which
appeared in the *Spectator* of January 18th,
and one or two of them suggest a doubt as
to the veracity of the story. I write, therefore, to tell you that it is literally true, only
that the policeman was away for his holiday
instead of having influenza, and the case
came off on Tuesday instead of Saturday.
My dog is a pug, a very choice specimen
of his kind, and was given to me by the late
Dr. Wakley, editor of the *Lancet*, who was
a great connoisseur in dogs. His intelligence
is really marvellous, and he has done many
things as extraordinary as the one related by
Miss Wood.

He is devotedly attached to my baby, and
always accompanies me in my morning visit
to the nursery. On one occasion the child
(who is just as fond of him as he is of her)
was very ill, and for three weeks was unconscious. As soon as this was the case,

14

the dog ceased to go near the nursery, as if by instinct he knew he would not be noticed. Mr. Walters from Reading was attending the baby, and the dog soon got to know the time he paid his visits. He would watch him upstairs, and when he came down listen most attentively to his report. At length the child was pronounced out of danger. The very next morning, up went master Sam, made his way straight to the child's cot, and stood on his hind legs to be caressed. Although she had taken no notice of any one for some time, she seemed to know the dog, and tried to move her hand towards him to be licked. He quite understood the action, licked the little hand lovingly, and then trotted contentedly away. After this he went up to see her regularly, as he had been accustomed to do. He is quite a character in the town, and nearly every one knows Sammy Weller.

Before I had this dog, I always thought I understood the difference between reason and instinct, but his intelligence has quite puzzled me.

MARY H. BARFORD.

ARE DOGS "COLOUR-BLIND"?

[*Jan.* 12, 1884.]

YOUR correspondent, "W. H. O'Shea," has found several dogs "colour-blind." If black is a colour, I can give several instances in which a black retriever dog of mine was certainly *not* "colour-blind." He had the greatest antipathy to sweeps and coalheavers, and would fly at them if not fastened up or carefully watched. He would even bark at a passing hearse! In all other respects, he was the best-tempered dog in the world, and I can only imagine that when very young he must have been ill-used by either a sweep or a coalheaver.

C. R. T.

LUCKY AND UNLUCKY.

[*April* 28, 1877.]

As letters telling of dogs and their doings occasionally appear in the *Spectator*, perhaps the following rather pathetic anecdote of a dog I know well may also find a place there. Two or three weeks ago, Lucky—so called from having, when an outcast, found its present happy home—perhaps by way of showing its gratitude to its benefactors, presented them with five small Luckys, or rather, with one exception, Unluckys, as the melancholy process always resorted to with these too-blooming families had to be carried out in this instance, and the five were reduced to one. Poor Lucky was inconsolable, looking everywhere for them, and looking, too, with such appealing eyes into the faces of her friends, and asking them so plainly where they were. Near her kennel was an inclosed piece of ground for pigeons, and as it was discovered that rats were carrying off the young pigeons, and as Lucky had carried off one or two rats, it was decided one night to leave the door of the pigeons' house open, that Lucky might have the run of it ; and the next morning, side by side with the puppy, was found a baby pigeon, looking quite bright and at home, but hungry, and poor Lucky, proud of the addition it had made to its family, was looking more contented than it had done since the loss of its puppies. The pigeon must have fallen from its nest, some distance from the ground, and Lucky, while on the look-out for rats, must have found it, and carefully carried it to her kennel, with the vague feeling, perhaps, that it was one of her own lost little ones "developing" a little curiously. Unfortunately the arrangement could not be a permanent one, and the famished little pigeon was put back into its own nest, to be found again the next morning in Lucky's bed, but this time dead. The old birds seem to have deserted it, and it had died of starvation. If Lucky could give this account herself, it might be much more interesting, for it was thought not at all improbable that she had actually rescued from a rat the bird she was so anxious to adopt, as a small wound was found upon it such as a rat might have made, and as a young pigeon had been taken the night before from the same nest; but this is only conjecture, and Lucky only could tell us the facts; how often it would be interesting, if our humble friends could tell us their adventures! A friend who is staying with me tells me that a few months ago her dog was lost for a week, and at the end of that time it came back one night in a scarlet ruff and spangles, and looking altogether dreadfully dissipated. Evidently it had been the "performing dog" in some show, "Punch and Judy" perhaps; being naturally a clever dog, it would quickly have learnt the part of "Toby" in that delightful and time-honoured exhibition. If it could only have written also an article entitled "A Week of My Life," with what pleasure the *Spectator* would have published it!

S.

THE COURAGE OF ANIMALS.

[*Feb.* 11, 1893.]

IN the *Spectator* of December 31st, which, although a regular subscriber to your valuable paper, I only happened to see to-day, owing to absence from home, I notice a reference in the article entitled " The Courage of Animals," to the fact that the wild dogs of India attack and destroy tigers. I have no personal knowledge of the matter, but I have been told by an Indian officer that the *modus operandi* of the "red dogs " is as follows :— Having found their tiger they proceed, not to attack him at once, as might be inferred from your article, but to starve him until they have materially reduced his strength. Night and day they form a cordon round the unfortunate beast, and allow him no chance of obtaining food or rest ; every time the tiger essays to break the circle, this is widened as the pack flies before him, only to be relentlessly narrowed again when the quarry is exhausted. After a certain period of this treatment the tiger falls a comparatively easy prey to his active and persevering enemies. This theory of their plan of attack, while it may detract somewhat from the wild dogs' reputation for courage, must add considerably to our estimate of their intelligence.

Edward Paul, Jun.

SOME FACTS OF MATERNAL IN-STINCT IN ANIMALS.

[*Oct.* 1, 1892.]

I LATELY met some friends who had with them a little dog, called Vic, who had adopted the family of a cat in the house, and, while in possession, would not let the mother come near her kittens. The kittens were kept in a very tall basket, and Vic would take them in her mouth, and jump out with them one by one, and then carry them into the garden and watch over them, carrying them back in the same way after a time ; at other times, lying contentedly with them in the basket. Of course Vic' had to be forcibly removed when the adopted family required their mother's attention for their sustenance. I also have met a friend who saw a hen-hawk, who was in a cage, mothering a young starling. Three young, unfledged starlings were given the hawk to eat. She ate two, and then broodled the other, and took the utmost care of it. Unhappily, the young starling died ; and from that moment the hawk would touch no food, but died herself in a few days.

The same friend was on a mountain one day, when a sheep came up to him, and unmistakably begged him to follow her going just in front, and continually looking round to see if he was following. The sheep led him at last to some rocks, where he found a lamb fast wedged in between two pieces of rock. He was able to liberate the lamb, to the evident joy of the mother.

I myself once saw a cat "broodling" and taking care of a very small chicken, which, being hatched first of a brood, had been brought into a cottage and placed in a basket near the fire. It managed to get out of the basket, and hopped up to the cat, who immediately adopted it.

WM. WALSHAM WAKEFIELD.

HAVE ANIMALS A FOREKNOW-LEDGE OF DEATH?

[*April* 30, 1892.]

IN a recent *Spectator* there is a quotation from Pierre Loti to the effect that "animals not only fear death, but fear it the more because they are aware that they have no future." Pierre Loti is a brilliant novelist, but I am not aware that he is a scientific naturalist, and I trust his idea is a mere chimera. Loti would take from the brutes the one privilege for which men may envy them, and endows them with a knowledge of the aftertime that we have only by revelation. However, two common-sense naturalists have published their belief that the lower animals have a foreknowledge of death, and one of them goes so far as to give an account of an old horse committing suicide. He says the animal frequently suffered from some internal disease, and that it deliberately walked into a pond, and, putting its nostrils under water, stood thus till it dropped dead from suffocation. The incident, I think, is easily explained.

Many horses drink in the manner described, and in old horses heart-disease is not uncommon. I imagine the stoppage of respiration caused a sudden and natural death from heart-disease.

I should like to ask naturalists who think animals know that they must die, where they draw the line. They must stop somewhere between a dog and a dormouse. Poets have made far more frequent allusion to the subject than naturalists, and they may be quoted on both sides. Philip James Bailey, in illustration of his contention that hope is universal, says: "and the poor hack that sinks down on the flints, upon whose eye the dust is settling, he hopes to die." But we have on the other hand Shelley's Skylark, with its "ignorance of pain," because it differs from men who "look before and after." Wordsworth's little girl of eight knew less than her dog, if she had one, for, says the poet, "what could she know of death?" I admit that when the carnivora have crushed their prey to death they cease to mangle them; but I fancy that is only because there is no more resistance ; and a bull will trample on a hat and leave it when it becomes a shapeless mass. The nearest thing I ever saw to an apparent foreknowledge of death, was in the case of that least intelligent of dogs, a greyhound. I had to shoot it to prevent useless suffering from disease. It followed me willingly, but when I led it to a pit prepared as its grave it instantly rushed off at its best speed. I suggest that it saw instinctively something unpleasant was about to happen, but it does not follow that death was present to its mind. Domestic poultry will furiously attack one of their number that struggles on the ground in its death-agony. They do not dream of death ; they think its contortions are a challenge to combat.

R. Scott Skirving.

OUR FOUR-FOOTED FRIENDS, BIG AND LITTLE.

[*Nov.* 8, 1873.]

MAY I be permitted to question, in the most friendly way, the assumption of "Lucy Field," in your last issue, that the lives of small dogs are in constant jeopardy from "a race of giant dogs, and exceptionally large dogs," at Muswell Hill? If it be so, then, surely the "giant dogs" of that region are exceptions. My experience goes to confirm the truth taught by Sir Edwin Landseer's "Dignity and Impudence," a fine print of which adorns my portfolio. I had a broken-haired friend, weight about eight pounds, learned in two languages, canine and English, who rejoiced in the name of Teens, given him by babes with whom he condescended to play, because he was a "tiny, teeny dog." I must confess that my late friend—alas! that I should say late—who was chivalrically brave in killing rats and carrying on war with cats, was a very bully, a kind of Ancient Pistol towards big dogs. To see him meet

a Newfoundland or large retriever was as
good as a play. Teens, with his tail curled
like the spring of an ancient watch, his
broken-haired back stiffened with indignation,
would stand and give the pass-word all dogs
seem to know, and be overhauled and
examined as he walked round the giant like
an English gunboat by a Spanish fifth-rate ;
but when once the enemy turned his back,
Teens exploded like a cracker, running
under the big dog's nose, and often springing
at his lip. His gigantic, but generous foe
(or friend) always fled, or walked away,
followed by a torrent of abusive barks,
which, from their peculiar intonation, I took
for dog-slang, and Teens returning with an
impudent smile on his countenance, wiped
his feet on the pavement as a sign of
triumph. I have seen him do this a hundred
times, and never saw a big dog attempt to
punish his impudence. Jeems, a black-and-
tan of smaller weight, who seemed to walk
upon springs, and who on work-a-days was
called Jim, and James on Sundays, which
day he perfectly well knew, was more like

Parolles. He bullied big dogs at a distance, and seldom stood up to them like the truculent Teens, and, although he ran away, was seldom pursued and never hurt, while the Claimant (he was for his size unwieldly in fatness as a pup), who (or which) still lives with me, is now bullying a shambling retriever pup, full-grown, but, like Cousin Feenix, uncertain as to his gait, who good-naturedly submits to it. Here, perhaps, there is danger ; for very big pups will pursue any little thing that runs away, and one of their large paws, which they put down as if they wore heavily clumped boots, might certainly crush the life—a very noisy, fussy, busy life it is—out of my small and impertinent, pretentious Tichborne. This dog, by the way, brings down his mistress her boots, as a hint for her to take a walk, and blows like a trumpet or young walrus under the door to be let in, having been corrected for scratching the panel. I end as I began, by assuring you that my experience, no less than that of my friends, lies in the direction of extreme generosity exhibited by large dogs towards small ones ; I would not deny that a large dog may now and then punish an impudent and aggressive toy-terrier, but, as a rule, we can only wonder at the providential wisdom which makes them so generous and forbearing ; having a giant's strength, they seldom indeed use it like a giant.

HAIN FRISWELL.

DOG CONSCIOUSNESS.

[*Nov.* 2, 1872.]

OUR terrier Crib took upon himself yesterday to add his testimony to your view of "dog-consciousness," as expressed in the *Spectator* of the 19th ult. Crib verges on perfection, save that he is frantically jealous of any other animal who may receive attention, but yesterday he rebelled against the injustice of being compelled to eat all his dinner, and refused to swallow one special piece of bread; but finding that his refusal was not accepted, apparently made a virtue of necessity, and gulped down the bread with a look and wag of the tail, giving me to understand that I ought to be satisfied, which I was not, as I observed a slight swelling in one cheek. So concealing my suspicion I furtively watched. Crib also occasionally eyed me, lying down and then walking round the room, and sniffing in the corners, as he is wont to do. In a few minutes, and when I appeared safely absorbed in my paper, he made his way slowly to where pussy was lapping her saucer of milk; passing her without stopping, he cleverly discharged the hated mouthful into pussy's milk, and continuing his walk to the rug, laid himself down and slept the sleep of the just.

C. S.

A DOG STORY.

[*June* 1, 1895.]

PERHAPS you will allow me to add another
to your interesting list of dog stories. In a
house where I once boarded there was a
large and remarkably sagacious St. Bernard
mastiff, who used to come into my sitting-
room and give me his company at dinner,
sitting on the floor beside my chair, with his
head on a level with the plates. His master,
however, fearing that he was being over-fed,
gave strict injunctions that this practice
should no longer be permitted. On the first
day of the prohibition the dog lay and sulked
in the kitchen ; but on the second day, when
the landlady brought in the dishes, he stole
in noiselessly close behind her, and while for
the moment she bent over the table, he
slipped promptly beneath it, and waited.
No sooner had she retired than he emerged
from his hiding-place, sat down in his usual
position, and winked in my face with a look
which seemed to say, "Haven't I done her!"
In due course, the good woman came to

change the plates, and as soon as he heard her step, he slunk once more under the table; but in an instant, ere she had time to open the door, he came out again, as if he had suddenly taken another thought, and threw himself down on the rug before the fire—to all appearance fast asleep. "Ah, Keeper; you there, you rascal!" exclaimed his mistress, in indignant surprise, as she caught sight of him. The dog opened his eyes, half raised his body, stretched himself out lazily at full length, gave a great yawn as if awakened from a good long sleep, and then, with a wag of his tail, went forward and tried to lick her hand. It was a capital piece of acting, and the air of perfect guilelessness was infinitely amusing.

GEO. McHARDY.

WOW: A STORY OF A CAT'S PAW.

[*March* 23. 1872.]

I THINK you will be interested in the following anecdote of a distinguished foreigner.
One of the happiest results of that abandon-
ment of their ancient exclusiveness which
has rendered us familiar with the Japanese,
has been the arrival on these shores of a
very pretty fluffy little dog, a born subject of
the Mikado, who hails or rather barks from
Nagasaki, and who is happily domiciled with
a friend of mine, of a sufficiently elevated
mind to esteem at its proper value the privilege of being the master of a clever and
refined dog. The child of the sun and
the earthquake has been named Wow, an
ingenious combination of the familiar utterance of his kind with the full-mouthed
terminals of the language of the merely
human inhabitants of his country. My own
impression is that Wow smacks rather of the
melodious monosyllabic tongue of the Flowery
Land than of that of the Dragon country;
but this is a detail, and, as a young naval

officer newly come from Nipon remarked to
me lately, with much fervour, " Thank God!
a fellow isn't obliged to learn their lingo."
Wow has made himself at home and happy
in his Northern residence with all the
courtesy and suavity of a true Japanese, and
has attached himself to his master with
apparent resignation to the absence of pigtail
and petticoat, articles of attire replaced in
this case by the wig and gown of a Q.C.
About this attachment there is, however,
none of the exclusiveness which characterises
the insular dog.   Wow is a politician, or at
least a diplomatist, and he desires to main-
tain friendly relations, with profitable results
to himself, with everybody.   He succeeds in
doing so to an extraordinary extent, of which
fact his master lately discovered evidence.
Very strict orders, including the absolute
prohibition of bones, had been issued with
regard to Wow's diet.   The ideas of a
country in which little dogs eat, but are not
eaten, require liberality in his opinion, and
Wow made up his mind he would have his
bones without incurring the penalties of dis-

obedience, which his master, in the interests
of the delicate foreigner, was determined to
inflict. A commodious and elegant residence
was fitted up in the study for Wow, and he
was permitted free access to the upper floors
of the house, but the line was drawn at the
kitchen staircase. That way lay bones and
ruin, and its easy descent was interdicted by
stern command, which Wow understood as
clearly as did its utterer, though he at first
affected a simple and unconscious misappre-
hension. Then Wow was reproved and
gently chastised, an administration of justice
performed with the utmost reluctance by
his master, but with the happiest results.
Nothing could be more admirable than
Wow's submission, more perfect than his
obedience. He never looked towards the
kitchen stairs, and would attend at the family
meals without following the retiring dishes
with a wistful gaze, or betraying a longing
for the forbidden bones by so much as a sniff.
Attached to the lower department of the
household is a humble cat, a faithful creature
in her way, but not cultivated by my friend

as I could wish. With this meek and useful animal Wow contracted a friendship regarded by his master as a proof of his amiability and condescension. (In my capacity of narrator I am compelled to use the latter somewhat injurious term—as a private individual with an undying recollection, I repudiate it). But the single-minded Q.C. had something to learn of the four-footed exile from the Far East concerning this intimacy. Coming into his study one day at an unusual hour, he saw the cat—I do not know her name, I am afraid she has not one—stealthily depositing a bone behind a curtain. Presently she went downstairs, and returned with a second bone, which she conveyed to the same place of concealment, whence proceeded a gentle rustling and whisking, suggestive of the presence of Wow, whose house, or pagoda, was empty. Then arose the Q.C., and cautiously peeped behind the curtain, where he beheld Wow and his humble friend amicably discussing their respective bones, Wow's being the bigger and the meatier of the two.

Thus did the Japanese exile illustrate the cosmopolitan story of the catspaw (with the improvement of making it pleasant for the cat), and accomplish the proverbially desirable feat of minding both his meat and his manners. If we could be secured against their imitation, it would be pleasant to ask our own domestic pets the problems :

" What do you think of that, my cat ? "
" What do you think of that, my dog ? "

A CONSTANT READER AND DISCIPLE.

THE BIOGRAPHY OF SPRIG.

[Jan. 20, 1872.]

I DARE not hope to equal the eloquent and most touching biography of Nero, with whom I had the honour of a slight acquaintance. But I was the possessor of an animal who, in his way as a dog, not a cat, for originality of character, reasoning power, talent, and devoted affection I have never seen equalled in his species, and you and your readers may possibly be interested by a sketch of his biography.

Where Sprig was born I do not know, nor had I any acquaintance with his parents. One morning several years ago I chanced to go down stairs early, and found the milk-boy at the hall door, delivering his daily supply to the cook. In the courtyard before my house was a bright-looking rough terrier of small size, frisking about very cheerfully, trying to catch the small stump of a tail which some cruel despoiler had left him. As he was engaged in this pastime, a large brown retriever entered the gate, to look on,

I suppose, for he had an amused expression
of face, and was wagging his tail amicably.
Sprig, however, though but a mite in com-
parison, decidedly resented the intrusion, and
flew at the retriever's throat, from which he
had to be choked off by his owner, who
brought him back in his arms. The
little fellow was in the highest state of
excitement and anger, his bright, intelligent
eyes flashing, and his hair bristling. He
was indeed most amusingly fierce, but was
soon calmed when he was shown, and told,
that his enemy had fled, whereupon the
following colloquy ensued between myself
and his owner. Myself: "And where did
you get that dog, boy? You did not steal
him, I hope?" Boy, in a rich Dublin
brogue : "Ah, now! would I stale anythin',
yer honner, an' me the poor milk-boy? Is
it stale him? Bedad, it's my father's cuzin
that's at the Curragh! Sure he's a corporal,
so he is. He brought him, and he sez,
'Yez'll get me a pound for him, and no less.'
So it's a pound I want for him, sur, and
nothin' less. An' sure John Lambert knows

me well—so he does!" When John, my
servant, was sent for, he gave a good account
of the lad, and as he entirely approved of
Sprig, I gave the sovereign, showing it to
the dog, whose wondering eyes were glancing from one to the other. Then I said to
the boy, " Put him into my arms, and tell
him he belongs to me;" and he did so.
The little fellow looked curiously and wistfully at the lad, who, to do him justice, had
tears in his eyes, and then nestled into my
breast, licking my hands and face. When
my daughter came down stairs, I took up
Sprig and placed him in my youngest
daughter's arms, a process he appeared to
comprehend perfectly, and told him she was
his mistress ; nor to the day of his death did
he ever falter in his devoted allegiance to
her. He was very fond of me and of us all,
but his deepest love was for his mistress, and
on many occasions was most affecting to see.
She was often delicate, and once had a sharp
attack of typhus fever. In this illness Sprig
never left her. He would lie at the foot of
her bed watching her, and would sometimes

creep gently up to her, put his paws round her neck, and lick her hands softly, while the pleading of his large eyes looking from his mistress, in her unconscious delirium, to her sister and me, was touching in the extreme. Indeed, there were then many sad illnesses, but Sprig was always the same. As my child grew stronger and better her little friend would amuse her by the hour together; sit up, beg, preach, play with his ball, and try in humble doggie fashion to beguile her of her pain. But I am anticipating.

Sprig was, I believe, what is called a Dandie Dinmont, and as he grew up he became, for his class, a very handsome, as he was a sturdy, little fellow, with great strength for his size. He was a reddish-brown colour, more dark-red than brown, like a squirrel, with white below, and a delightfully fuzzy head, and a breast of long soft white hair. His eyes were that peculiar bright liquid "dog" brown which is capable of so much expression, and he grew to have a long moustache and beard. Even the most un-observant of dogs admired him, for he resembled no terrier I have ever seen. I think he would have won the prize of his class at the Dublin Dog Show, had it not been for a terrible accident he met with in being wounded by a large foxhound in a neighbouring orchard. His neck was then torn open, and he was rescued by John only in time to prevent his being killed. As it was, it was weeks before he could walk— and how patient he was all the time! and as the wound healed it left a thickening of his skin which had an awkward look. Sprig was, however, "highly commended." In his youth he was perhaps rather short in his temper, and always resented in the most distinct manner any liberty that was taken with him. To tread upon his foot was perilous, but he was at once pacified if an apology was made that it was accidental; but to pull his tail wilfully was an insult which he resented bitterly, and for which much atonement was necessary, or he would go under the sofa and cry in his peculiar manner when offended.

As he grew up, Sprig developed various

talents which were highly cultivated. His
greatest pleasure, perhaps, was in an india
rubber ball, with which his gambols were
indescribably pretty and constant. It was a
great distress when he lost or mislaid his
ball, and he was miserable till he found it,
or another was brought him. It was a cruel
thing to say, when one of us went to town,
" Sprig, I will bring you a new ball," and as
sometimes happened, to forget to do so. On
return he would sniff about the person who
had gone, poke his nose into his or her
pockets, and if disappointed could hardly be
soothed, but would go away and have his
quiet cry to himself. Sometimes a kind
friend who knew him might bring him a new
ball ; but it very much depended on who
presented it whether it was accepted or not,
and I am afraid that too frequently for his
good manners he turned it over contemp-
tuously with his nose and left it for the old
one, which, gnawed, bitten, and broken, was
still the favourite. I used sometimes to
make a ball squeak by pressing the hole
against my hand, and I believe he thought

it was in pain, for he would whine piteously, and would not let me rest till he had it again in his possession. It was most amusing to see him when a parcel of new balls arrived, he having been told beforehand that one was coming. He would find out directly who had it, and become impatient and cross indeed if he did not get it directly. When the parcel was given him, his great delight was to open it himself and select *one*. A red ball was usually preferred, but not always. All were subjected to the most varied trials —gnawed, smelt, and rolled, till the one which pleased his fancy was finally selected; of the rest he would take no notice whatever.

Sprig was thoroughly a gentleman, and on most occasions he was most attentive to lady visitors. He never noticed gentlemen. On one occasion, when my daughters were out, a dear friend called (Nero's mistress). She told us afterwards that Sprig had been a most attentive beau. He met her at the hall door, welcomed her in his odd fashion, trotted before her into the drawing-room, looking behind him to see if she followed. He then jumped upon the ottoman, inviting her to sit down ; when she was seated he brought his ball and went through all his tricks with it, sat up on his hind legs, begged with his paws, preached to her in his own queer way, and kept her amused till, no longer able to remain, she bid him good morning and left, evidently to his disgust. " Could he have spoken," she said afterwards, " he would have told me to wait, for his mistresses would soon be back ; the look was in his face, but the words were wanting." His attention to visitors was never omitted. When we had a ball or evening party, he would await, with John Lambert, the several arrivals at the hall door, welcome each new party, and usher them in a solemn manner into the drawing-room or tea-room, returning for a new set to his former place. Nor did he want for an occasional cake or biscuit at the tea-table ; " he was so amiable," said the young ladies, " he could not be resisted."

As an instance of how perfectly he understood what was said to him, I may relate

that one hot day I had walked out from
town, and being thirsty went into the dining-
room for a drink of water. I saw Sprig's
ball under the table, and when I went into
the garden where my girls were sitting they
said, "Sprig has lost his ball, and is perfectly
miserable." After I had sent him to look
about for it, I said, "Now, Sprig, I know
where it is; I saw it in the dining-room
under the table; go fetch it." He looked
brightly at me, and I repeated what I had
said. He trotted off, and while we were
wondering whether he had understood me,
he returned with it in his mouth quite
delighted. I have mentioned his preaching,
which may sound rather irreverent, but it
was an accomplishment entirely of his own
invention. When seated in a chair after
dinner, and requested to preach, he would
sit up, place his forepaws gravely on the
table, and then lifting up one paw as high
as his head, and then the other, deliver a
discourse to the company in a sort of gur-
gling, growling manner, with an occasional
low bark, which was indescribably ludicrous

to see and hear. What he meant by it we could never find out, but I question whether he prized any of his accomplishments more than this.

Sometimes, but not often, he would go out by himself to take a walk, we supposed to see his friends, for I never heard that he had any love affairs. If we all, or my daughters, or myself, met him on his return, I, or they, or we all might call to him, notice him as he brushed past us, or ask him to come for a walk. No. He would have none of our company ; he would cut us dead, and go toddling home, his tail more erect and quivering than ever ; never hastening his sedate pace, and giving his usual kick-out with one hind leg every third or fourth step, as was his custom. He would have no connection with us ; that was quite clear and decided. Sprig was very fond, too, of a walk with his mistresses or with me, and, though never taught it, would always wipe his feet clean on the hall mat as he came in. I am now going to relate an anecdote of Sprig which I know is almost beyond credibility, but the

occurrence so displayed his power of thought
and reason that I cannot withhold it. My
usual haunt is my den, as I call it, a large
room at one end of our old rambling house.
There Sprig never came unless with his
mistresses, and indeed never was easy when
he was there. I had begun a large full-
length picture of my daughters, and Sprig
and Whisky, a small Skye puppy, were to be
painted lying at their feet. As the picture
progressed, Sprig seemed to understand all
about it, and paid me the compliment ot
wagging his tail at the portraits. One day
my girls had been sitting to me, and it was
now Sprig's turn to sit. I put him into the
proper position and told him to lie still, and
he proved a most patient sitter. When the
sketch of him was finished, I showed it to
him ; I think he was pleased with his likeness, for he licked my face ; but as he smelt
at his portrait, he did not like himself, and
growled. Whisky was now put into position,
but was very restless, although Sprig scolded
her by snarling at her. Next day I had put
the picture against the wall near the window,

and before a few steps which led up into my bedroom, and was busy perched on a step-ladder with the after-portion of it. By and by I heard a great scratching at my bedroom door, which was closed, and Sprig whining to get in. I thought this odd, but it was too much trouble to come down from my perch, and I told him to go away. He, however, only whined and scratched the more. I therefore descended, and getting behind the picture, went up the steps and opened the door. Sprig did not notice me, but pushing past me hurried down the steps, and then, as I emerged into the room, looked up to me blandly, and actually sat down in the place in which I had put him the day before. I said to him gravely, though infinitely amused, " No, Sprig, I don't want you to-day ; look, the colour is all wet, go away to your mistress." He looked very blank and greatly disappointed, and stood up with his tail drooped. Suddenly a bright thought seemed to strike him, as if he had said, " Now I have it !" Whisky had got hold of one of my slippers, and was playing with it in my bedroom, and Sprig, rushing up the steps, seized her by the "scruff" of her neck, dragged her howling down the steps, and put her, I can use no other words, into the place where she had been the day before. He then came to me frisking about, and could he but have spoken, would have said, " If you don't want me, you must her, and there she is!" He was quite triumphant about it; and dirty as I was, and palette in hand, I took him forthwith to the drawing-room and told them what had happened.

I could tell numberless other stories of the reasoning power and intelligence of our little pet, but I should trespass at too great length on your patience. I could describe a curious friendship which sprang up between him and a German friend who was staying some time with us; how he learned many new tricks from him, and was taught to hop on his hind legs from one end of the drawing-room to the other, with our friend hopping backwards before him; I could describe his evening romps with my dear father, never omitted while my father lived; and the many curious traits by which his great love for us was perpetually displayed—how he learned to crack nuts of all kinds, and to pick out the kernels like a squirrel—how he never went into the servants' hall or the kitchen, and refused to associate with the servants, though friendly with them, and especially with John Lambert, his fast friend. But I must bring this sketch to a close.

We had been absent about a year in Germany and the South of France. After we left, Sprig was inconsolable, and would not eat; but the cook made him little curries and rice, and after a time he became more resigned. We only heard that he was well, and hoped we should find him so. The day we arrived I thought he would have died for joy. He gasped for breath, and lay down, and when taken up by his mistress lay in her arms almost insensible. It was long before he came to himself, and when he did revive, it is quite impossible to describe his delight, or what he did. He was, indeed, quite beside himself with joy, scouring about, dragging his mistress here and there, doing

all his tricks in a confused manner, and, in
short, behaving after a very insane fashion
indeed. We noticed he had a slight cough ;
but he seemed otherwise quite well, and we
thought it would go away , but it increased,
·and at that time there was an epidemic of
bronchitis among dogs. We sent him to an
eminent veterinary surgeon, who blistered
him (and how patient the poor fellow was
under the pain cannot be told), but though ·
relieved for the time, the end was near.
One morning he was seen to do an apparently quite unaccountable thing. He
took his son Terry (whom he was never
known to notice except by knocking him
over and standing upon him, growling
fiercely), all round our village, and visited all
the dogs in it. John saw him doing this
early in the morning, and told me of it. I
suppose he was commending Terry to their
favour. He coughed a great deal all day,
and breathed heavily ; but in the evening he
was very bright, and to all appearance much
better, and insisted on doing all his tricks
till it was time to go to bed. Sprig never

would go to bed willingly. John used to
come to the drawing-room door and call him,
and he would go to it, but stand growling
till he was caught up and carried off. That
evening, as we remembered, he seemed more
than ever unwilling to go, but was caught up
and carried away.

In the morning, about six o'clock—it was
summer-time—I was just about to get up,
when John Lambert knocked at my door,
and came in with Sprig in his arms. He did
not speak, and I asked him whether Sprig
was worse. "He's dead, sir," said he, with
the tears rolling down his face, and hardly
able to speak. "Quite dead, sir; he must
have died only a little while ago, for when I
went to let him out, I found him dead and
quite warm, as he is still." I am not
ashamed to write that my eyes felt very
blind, but there was no hope; the dear little
fellow was quite dead; he had died calmly,
and his eyes were bright; they had not
glazed.

We buried him, John and myself, when he
was quite cold and stiff, by a rose-tree at the

end of the garden. Poor John could hardly dig the grave, and his tears fell fast and silently and upon dear old Sprig as we covered him up for ever. I wish I could write a fitting epitaph for a creature who, through his life, was a constant source of pleasure to all who knew him.

M. T.

A DOG STORY.

[*June* 8, 1895.]

A FRIEND thinks I ought to add to the collection of dog stories appearing in the *Spectator*, one which is within my own knowledge, and may appear deserving of publication. My uncle, a well-known Chairman of the Bench of Magistrates in a western county, had a tenant on his estates who occupied a farm not far from the River Severn. The farmer possessed a favourite dog, who slept at the foot of his bed every night. When a brother emigrated to Canada, the farmer gave him the dog as a travelling companion. In the course of time the news arrived that the emigrant and his family, together with the dog, had safely reached their destination—a farm in the interior of Canada some days' journey from the port where they landed. At a later date the brother in Canada wrote to his family in England saying that the dog had disappeared. Some time afterwards the dog came back to the farm of his old master about three miles from Gloucester, and though at first it could hardly be believed that he was returned from Canada, yet he soon established his identity by taking his old place at the foot of his master's bed at night.   Inquiries were made, and the dog's course was traced backwards to the River Severn, thence to Bristol, and thence to a port in Canada.   It appeared that, after running from his home in Canada to the seaport, he selected there a vessel bound for Bristol, and shipped on board.   After arriving at the Bristol basin, he found out a local vessel trading up and down the River Severn (locally called a " trow "), and transferred himself to her deck.   When he reached the neighbourhood of Gloucester, the dog must have jumped into the Severn and reached the shore nearest to his old home.

I can vouch for the truth of this story, from information received from my relations on the spot shortly after the occurrence took place.   I knew the farm well, and the farmer who occupied it.

H. C. N.

A CAT-AND-DOG FRIENDSHIP.

[*June* 8, 1895.]

THE interesting letter, " A Canine Nurse," in the *Spectator* of May 18th, recalls to mind an equally curious event in cat and dog life which occurred some years since in a house where I was living, but with the additional interest of a hen being also implicated.

In the back-kitchen premises of an old manor-house, amongst hampers, and such like odds and ends, a cat had a litter of kittens. They were all removed but one, and as the mother was frequently absent, a hen began laying in a hamper close by. For a time all things went well, the hen sitting on her eggs and the cat nursing the kitten within a few inches of each other. The brood were hatched out, and almost at the same time the old cat disappeared. The chickens were allowed to run about on the floor for sake of the warmth from a neighbouring chimney, and the kitten was fed with a saucer of milk, &c., in the same place,

both feeding together frequently out of the
same dish.   The hen used to try to induce
the kitten to eat meal like the chicks, calling
to it and depositing pieces under its nose in
the most amusing way ; finally doing all in
its power to induce the kitten to come, like
her chicks, under her wings.   The result was
nothing but a series of squalls from the
kitten, which led to its being promoted from
the back to the front kitchen, where it was
reared until it was grown up.   At this time a
young terrier was introduced into the circle,
and after many back-risings and bad language
on pussy's part, they settled down amicably
and romped about the floor in fine style.
Eventually the terrier became an inveterate
rabbit-poacher—killing young rabbits and
bringing them home—a proceeding to which
the cat gave an intelligent curiosity, then a
passive and purring approval, and finally her
own instincts having asserted themselves,
she went off with the dog, hunting in the
woods.   Our own keeper reported them as
getting " simply owdacious," being found a
great distance from the house ; and keepers

of adjacent places also said the pair were constantly seen hunting hedgerows on their beats. On one occasion I saw them myself hunting a short hedge systematically, the dog on one side, the cat on the other; and on coming near an open gateway a hare was put out of her form, and bounding through the open gate, was soon off; the dog followed, till he came through the gateway, where he stood looking after the hare; and the cat joining him, they apparently decided it was too big or too fast to be successfully chased, so resumed the hedge-hunting, each taking its own side as before.

They frequently returned home covered with mud, and pussy's claws with fur, and would lie together in front of the fire; the cat often grooming down the dog, licking him and rubbing him dry, and the dog getting up and turning over the ungroomed side to be finished. This curious friendship went on for six months or more, till the dog had to be kept in durance vile to save him from traps and destruction, the cat, nothing daunted, going on with her poaching until one day she met her fate in a trap, and so brought her course to an end. The dog was a well-bred fox-terrier, and the cat a tabby of nothing beyond ordinary characteristics, save in her early life having been fostered by a hen, and in her prime the staunch friend and comrade of poor old Foxie, the terrier. If there are " happy hunting-grounds ", for the animals hereafter, and such things are allowed in them, no doubt they will renew their intimacy, if not their poaching forays, together there.

R. J. Graham Simmonds.

THE SENSE OF BOUNDARY IN DOGS.

[*March* 14, 1885.]

I HAVE been much interested in the com-
munications which have appeared from time
to time in the *Spectator* in reference to
" animal intelligence." Recently my attention has been called to a somewhat striking
illustration of it, in the case of my own
dog and his canine neighbour next door.
Wallace is an Irish staghound, and is about a
year old. My neighbour's dog is a pointer,
and is considerably advanced in life. There
is no hedge nor fence separating the two
estates. The dividing line runs between two
stone posts about a foot in height, and more
than two hundred feet from each other. The
dogs have never been friendly, the pointer
having repeatedly driven Wallace back over
the boundary when he has caught him
trespassing. Both dogs, even when going at
full speed, stop the moment my dog has

crossed the line. How does the pointer know where the line runs, and how does Wallace know when he is safely across it ?

F. TUCKERMAN.

ADDITIONAL STORIES.

DOGS AND HUMAN SPEECH.

THE DOG AND THE HOT BOTTLE.

[*Oct.* 26, 1895.]

THE following example of canine intelligence may interest your readers, and help to establish the fact that dogs do understand human language more than is generally realised. Not long ago, one of my guests was describing to me one evening, after dinner, how much she suffered from cold feet, especially at night. In the course of our talk I said to her, " You ought to use a hot-water bottle, and I can lend you one to-night." On which she told me that she always took one about with her. In a very short time, my collie, having slipped from the room unobserved, returned with my friend's indiarubber hot-water bottle, which he had brought down from her bedroom. I inquired where it was kept, and was told it had been hung on a hook by the window.

So the dog must have taken some trouble to accomplish his purpose. I should add that the dog has a trick of bringing down shoes occasionally from upstairs, but has never before or since brought down any other article.

S. B.

THE DOGS THAT SHOWED WHERE THE KITTEN WAS HIDDEN.

[*Nov.* 30, 1895.]

I HAVE two dogs, a spaniel and a little Highland terrier, also a cat. The latter has a kitten, born last Monday week. All the rest of her family were drowned, and this, I suppose, has made her rather suspicious of being moved about, for on Saturday last her hamper was put out into the yard while the floor of the washhouse was scrubbed. It was put back again in the usual place, and the cat seemed quite happy. However, some hours after, the kitten was found to be missing, and the cat was sitting contentedly on a chair in the little hall. We all hunted high and low for the kitten, but could not find it. At last I returned to the dining-room, where the two dogs were lying before the fire, and I said casually to the terrier, "Do show me where the kitten is," never really thinking that she understood me, when she solemnly got up, walked round me, under the table, and came to my other side, then stood looking at a small cupboard, wagging her tail. I opened the cupboard, and there lay the kitten on a tea-cosy! I at once called to my cousin, who had by this time given up the hunt and was in her own room. She called to know where it was found, and I said, "Go down to the dining-room and ask the dogs to show you." She then went and said, "Dear dogs, do show me where the kitty is," and immediately the spaniel got up and went to the cupboard, looking at the door and wagging her tail. They certainly both understood what was wanted of them. The spaniel was born in 1887, and has been in my possession since she was about six weeks old. The terrier is about the same age, but I have only had her since December, 1890.

THE OWNER.

THE DOG THAT HEARD HE DID NOT GIVE SATISFACTION.

[*Nov.* 30, 1895.]

ABOUT a fortnight ago I was given a fox-terrier, on condition that if it did not suit me I should return it to the donor. Last Sunday evening I was sitting in the drawing-room with my wife, the dog lying on his mat by the fire. I said that I was dissatisfied with the dog, and should write and offer to return him. My wife urged me to do so then and there, and, after discussing the matter for a short time, I got up to pen the letter. As I did so, the servant came to take the dog for a run prior to turning in for the night. No sooner was the garden-door opened than off went the dog, full speed, into the darkness, and has not been heard of since. He had always been taken out in the same way before, and had always come in on being called. Whether he understood the conversation I cannot tell. All I can say is that I can offer no other explanation for his disappearance. My wife and the servant who let the dog out can vouch for the truth of these particulars. The letter which I wrote offering to return him lies before me unposted, " to witness if I lie."

G. S. LAYARD.

JOE AND THE TENNIS TOUR-NAMENT

[*Dec.* 14, 1895.]

MAY I add my testimony to the intelligence of dogs in the matter of understanding what is said in their hearing ? Several years ago I had a beloved mongrel fox-terrier named Joe. We were staying some months at Penzance, and the dog went everywhere with us, and knew the place well. One day we were, as usual in the afternoon, on the club tennis ground, when the secretary came up and warned me that on the following day, as there was to be a tournament, no dogs would be admitted to the enclosure. I promised to shut Joe up at home. That evening we missed the dog, and in the morning also he was not to be seen. When we went to look on at the tournament in the afternoon, we found Joe waiting for us; the groundman told us that the dog had been there all night, and would not allow himself to be caught. He had never slept out before, and he certainly must have

understood what was said.   We often used
to say, "We will drive to such a place
to-day, but Joe must stay at home," and
almost invariably, in whatever direction it
might be, before we had driven a mile, we
found Joe waiting for us by the roadside ;
he always grinned when we came up with
him.

HENRIETTA M. BATSON.

DOGS AND THEIR POWER TO FEEL EMOTION.

THE EMOTION OF GRIEF IN DOGS.

[June 22, 1895.]

YOUR article on " The Emotion of Grief in Animals," in the *Spectator* of June 15, leads me to send you an account of what happened to me. Some years ago I was out riding, accompanied by my two dogs—an Irish water-spaniel and a bull-terrier. I had a fall and broke my thigh. The distress of the dogs was touching to see. They ran to and fro, barking and howling, apparently to attract attention. When assistance came, I was carried home on a hurdle, the two dogs trotting one on either side of it ; and when the bearers put the burden down to rest, they jumped on to it, licking my face and hands. For several days the spaniel lay for hours in the carriage-drive, apparently watching for his master. One morning, when the postman delivered the letters, the servant gave the dog my newspaper, and with, " Bring it along, Paddy," he carried it upstairs into my room.   His joy at seeing me was worth beholding ; and from that day he regularly met the postman, carried the newspaper off, and laid it on my bed.   He was scarcely ever after absent from the room or the passage leading to it.

T. W.

FUNERAL OFFERINGS BY A DOG.

[*June* 29, 1895.]

THE dog story told by a correspondent in the *Spectator* of June 22, illustrative of "the emotion of grief in animals," recalls to my mind an incident in which a dog's grief at the loss of a companion, and memory, are both displayed.   Dutch was a brown retriever of advanced years ; Curly was reputed to be a Scotch terrier, but his appearance suggested some uncertainty in his descent.   Dutch was chained to her kennel, and Curly, who en-joyed his liberty, evinced his friendship by frequently taking bones and other canine delicacies to his less fortunate friend.   One morning Curly presented himself at the house evincing unmistakable signs of grief by his demeanour and his whines.   A visit to the kennel, where poor Dutch was found lying dead, showed the occasion of Curly's unhappiness.   We buried Dutch decorously under a vine in the garden, and supposed that Curly would forget the incident, but we were touched to see him, in the capacity of faithful mourner, frequently revisit the spot where his old friend was laid, taking with him by way of offering, choice bones, which he carefully buried by the grave. This practice Curly continued for two years, when we left the house.

A. E. W.

HOW WRINKLE MOURNED FOR DUCHESS.

[*June* 29, 1895.]

YOUR very interesting paper, in the *Spectator* of June 15, on " The Emotion of Grief in Animals," leads me to write to you upon what appears to be a very strong appearance of it in a pug-dog, who in many ways shows signs of almost human intelligence, thought, and judgment. Wrinkle was unusually strong and active for one of his race. Duchess, his canine friend and companion, nearly of his own age, was brought up with him, and was a large St. Bernard. These dogs always acted together ; Wrinkle did the thinking, Duchess followed his lead in everything, the smaller dog being fully accepted as the master. Among their amusements were mimic fights on the lawn, in which Wrinkle developed marvellous skill, and in races, which, by cleverness rather than speed, he generally won. Fierce as these mock-battles were, no case of a real quarrel ever occurred. They would share a bone amicably, Wrinkle taking always the first turn at it. After the dogs attained maturity, their play and companionship continued. It happened, unfortunately, some years ago, that the St. Bernard died by accidental poisoning. Wrinkle attended the funeral almost in silence, the only evidence of sympathy being the tears that ran down his short nose.

The successor of Duchess was a deer-hound, Huldah, a cheerful, playful, and gentle-tempered beast. Wrinkle accepted the new companionship complacently, did not resent an occasional occupation of his bed, and to a certain extent trained the deer-hound to assume the guardianship that the St. Bernard had always taken in their walks and excursions ; but play and romping were resented, the mimic fights and races were over for ever. A few days later a sprightly fox-terrier was added to the family, and received toleration and countenance to nearly the same extent as the deerhound, but play and sport were still refused. Wrinkle is sociable and friendly, in a dignified and superior way, with most of the dogs he meets, but has never been known to play with any since Duchess died; he insists on retaining his mastership, and seems able to assert it without ill-temper or quarrelling. To his mistress, my daughter, he devotes all his affection as he always did, but for none of his own race can he afford to give such love as he had for his lost friend.

E. W. Cox.

THE DOG AND HIS MASTER'S GRAVE.

[*June* 22, 1895.]

MAY I give another instance of a dog's fidelity to a dead master? The curate of a parish adjoining mine in the Vale of Evesham, having died in the hamlet in which he served, was buried in the parish church-yard, some two miles distant. His dog had had been shut up till after the funeral, and, when let loose, was supposed to be lost. It was found some days afterwards lying on its master's grave. He came from Newfoundland, and I rather think had brought the dog from thence. When I was dining with another incumbent near Evesham, *his* dog walked in. It had been given to a gentleman who lived near Birming-ham, and sent thither by train, but found its way back, more than thirty miles. The same thing happened, not long ago, near this, and the dog, which came from Londonderry, must have made its way all round Lough Swilly, a distance of many miles. It had been sent by railway and steamboat.

N. S. BATT, A.M.

"GREYFRIARS' BOBBY."

[*June* 22, 1895.]

IT must be a quarter of a century since
"Greyfriars' Bobby" blazed the comet of a
season. The authorised version of the story
is practically that which appeared in the
Spectator of June 15. If the question is not
raised now, it will be too late to do so
in the future. Was Bobby an impostor?
I have heard his achievements questioned
in Edinburgh. I have been informed that
Bobby was so trained in hypocrisy that
he lost all self-respect. The dog, it was
averred, went home with the sexton regularly at night, and returned with him to the
graveyard in the morning, and then, like any
other trained mendicant, took up his pitch on
the grave of his quondam master. Trained
or not, Bobby was an interesting little
fellow, and until his death, he was to be seen
by day on his master's grave, which he would
leave about one o'clock. Then he regularly
paid a visit to Trail's dining-rooms, con-
tiguous to the churchyard, where he was sure

of a hearty welcome, and having appeased his hunger, he would again hie away to the grave, receive visitors while the sexton received tips, and at nightfall leave the grave-yard with the grave-digger. If Bobby was an impostor, his career ought to be laid bare.

X.

[We do not believe in this view of "Greyfriars' Bobby," having received a totally different account of him in Edinburgh eight or nine years ago.—ED. *Spectator*.

A DOG WITH INJURED FEELINGS.

[*August* 17, 1895.]

MAY I send you another dog story? My
dog, a half-retriever, half-setter, has been
with me for six years since I rescued him as
a puppy with a can on his tail. He has followed me constantly, and though always very
friendly with everybody, has been devoted to
me both indoors and out. Lately a change
has come over him; he would come into my
room when called, but would take the first
opportunity to go out. He seemed to be
dull, to have lost his old joyousness in our
companionship. Last fall my children went
to England, and I thought he missed them.
He would leave my room to lie under the
kitchen-table, and would follow the hired boy
about the place, so I told the housekeeper to
keep him out of the kitchen, and the boy to
take no notice of him. It made no difference.
Forbidden the kitchen, he would leave my
room and lie in the hall. He had always
been accustomed to follow me almost every-

where, whether riding or driving ; but this year, thinking the journey to town (sixteen miles) and back too much for him, I had left him at the ranch when going to town. Last Saturday I was driving to town, the dog started to follow, and as the boy was going to send him back, I said, " Oh ! never mind; let him come," and he came with us. Now the whole mystery is explained. On our return, the dog quite resumed his old habits. The change was extraordinary. He comes into my room and stays there as a matter of course; he greets me every morning on coming downstairs ; he jumps round in the old joyous fashion when I go out—in fact, is himself again. Evidently the trip to town was one of his most cherished privileges, and he took his own way to show that he had no use for a master who deprived him of it.

L. C. H.

DOG FRIENDS.

[*December* 14, 1895.]

As I know your columns are always open to
well-authenticated stories of the wonderful
gifts of our four-footed friends, I venture to
think that you will be interested in the following anecdote. Thirty years ago I was
living in St. George's Square, Pimlico, and
near me—in Denbigh Street, at a distance
of ten minutes' walk—resided a well-known
journalist, Mr. Percy Gregg. He had a little
black-and-tan dog, for which I found a home
when his master was about to leave London.
It was reported to me that Jimmie always
left my house after breakfast. At first some
alarm was felt that he would stray; but as he
invariably returned after an hour's stroll, I
took him to be one of those "vagrom"
animals who cannot live without a prowl in
the streets, and I felt no anxiety. But I
ascertained that whenever he went away, he
carried off a bone or something edible with
him. I watched him one or two mornings,
and saw him squeeze through the area-railings,

on each occasion carrying a big bone, which he had great difficulty in steering through the iron bars. Being curious about the destination of the food I made up my mind to follow him. I tracked him to an empty house, next to that in which his former owner had lived. In a cellar in the area there lived a half-starved, ownerless terrier, who, I suppose, had once been a friend of Jimmie's, and whom my dog, in his days of prosperity, never forgot. Regularly the good little fellow trotted off to the empty cellar, and divided his morning's meal with his poor friend. The story is told of the great Napoleon riding over one of his battlefields—I don't know whether it was Wagram or Austerlitz—and pointing to a faithful dog watching the body of his dead master, with the words, " That dog teaches us all a lesson of humanity!" So did Jimmie.

THOMAS HAMBER.

BOB, THE AUSTRALIAN RAILWAY DOG.

BOB, THE RAILWAY DOG.

[*August* 24, 1895.]

I OFTEN see interesting letters in the *Spectator* about dogs, and I thought perhaps your readers might care to hear about the best-known dog in Australia, and his curious mode of life. His name is Railway Bob, and he passes his whole existence on the train, his favourite seat being on the top of the coalbox. In this way he has travelled many thousands of miles, going over all the lines in South Australia. He is well known in Victoria, frequently seen in Sydney, and has been up as far as Brisbane! The most curious part of his conduct is that he has no master, but every engine-driver is his friend. At night he follows home his engine-driver of the day, never leaving him, or letting him out of his sight until they are back in the railway-station in the morning, when he starts off on

19

another of his ceaseless journeyings. I have not seen him on our line for some time ; but noticed with regret last time he was in the station that he was showing signs of age, and limping as he walked.

ADELAIDE E. CRESWELL.

DEATH OF BOB, THE SOUTH AUSTRALIAN RAILWAY DOG.

[*Sept.* 21, 1895.]

KNOWING your constant sympathy with the canine race, I venture to enclose some extracts from the *Adelaide Observer* concerning a well-known character in the Colony.

CATHERINE E. BUXTON.

———

"It is but seldom that we feel called upon to record the death of a member of the canine family, but the demise of Bob, the well-known railway dog, will be mourned by many of our rising youth, and evoke a sigh from the travelling public and railway employés, among whom Bob was a great favourite. It was customary for Bob, whilst spending a few days in the city, to pay frequent visits to Mr. Evans's butcher shop in Hindley-street for meals. On Monday afternoon he was given his third meal by Mr. F. J. Preston, an employé of Mr. Evans, when shortly afterwards, about 3.10 p.m., he barked at a passing dog, and then, with a pitiful whine, fell dead. He was about seventeen years of age, and had only a few days ago returned from a trip to Broken Hill. Mr. L. M. Tier has claimed the body of the dog, and Mr. Nathan, in accordance with a promise made some months ago, will stuff it. A correspondent some time ago wrote the following interesting particulars about Bob's life:—'There is only one privileged individual in the province permitted at all times to use the Government railways without payment, and, further, without a pass. Even the late Chairman Smith has been asked for his ticket, and the importunate porter would take no excuse; but 'franked' on all lines, and on engine, in van, or carriage alike, the one constant traveller, who acts as though he believed the railways were made for him, is our hero. You may meet him to-day on the Serviceton line, and next week at Oodnadatia. He is well known in the Adelaide Station, and his friendly salute is often heard from the open window of a carriage on the Port line, as he enjoys a suburban trip. He is always welcome in the porters'-room, but

his favourite place is on a Yankee engine;
the big whistle and belching smokestack
seem to have an irresistible attraction for
him. His acquaintances on all lines are
numerous, and he often engages in such
lengthy salutations that the train by which
he has been travelling starts without him;
but he is never left behind, as he has a
perfect knowledge of how to mount a train
in motion. He is not particular as to how
far he goes in any given direction. He has
set out for a hundred-mile trip, but suddenly
changed his mind and also his engine at a
roadside station, and come straight back
again. He lives on the fat of the land, and
he is not particular from whom he accepts
his dinner. All the members of the staff
contribute willingly to his needs, and he
reciprocates these good offices by refusing to
reply to any appeals from the ordinary
public. It is very clearly established that
his sympathies are with the railway men,
though he is not on the committee of the
mission.' "

" I had the honour of the acquaintance of Bob, the railway dog, and I must say that he was one of Nature's canine gentlemen," writes Hugh Kalyptus, " always self-possessed, dignified without hauteur, friendly without being familiar, and courteous, inasmuch as he would always rise when addressed, pay attention to what was said to him, and never treat anyone superciliously, as I have seen many bipeds do. Bob made no difference between fustian and broadcloth. He was what I call a well-balanced Democrat, making no invidious distinctions, but treating all classes with courtesy, born of a correctly cast character. I have seldom seen a man with a ·more marked character than Bob. Although a notability, he never seemed conscious of it, but would walk the platform of a station anywhere between here and the end of the railway system in a calm self-contained style, like a person who had travelled much, accepting the greetings of his various friends as with the air of an equal, and it mattered not to him whether a lord, a statesman, or a

mere member of the mob patted his head,
he wagged his tail and walked on his wise
way. Bob had a capital memory, and
woe betide the person who treated him dis-
courteously—he would cut him dead the
next time. On one occasion an official
employed on one of the stations of the
Northern line, being a little lax in the liver,
had the presumption to kick Bob out of
his way as he lay sunning himself on the
platform waiting for a train. Bob never
got out at that station again. He cut the
station and its official dead ; and, if he had a
legacy to leave, it would not be that man's
name that would be mentioned in Bob's
will. I remember once in the course of a
several-hundred-mile bicycle trip I struck a
wayside station, and was entertained by
Bob with all the cordiality with which a
gentlemanly dog of confirmed character
greets one whom he knows to be a firm
friend of his race. He took a great interest
in my faithful ' Tyler ' bicycle, and, sitting
down at my side, sedately watched every
detail of the cleaning up, oiling, and other

incidental operations. The work appeared to secure his approval, and he gravely walked round the machine three times, examining all the parts, and, as nearly as a dog could, said, ' That's all right ; she'll do now,' and he politely accompanied me to the ticket-office, watched the booking process, and saw the bicycle safely disposed in the van. I thought it very kind and attentive of him ; he had evidently often seen the engine-drivers cleaning up their engines, and regarded my performance as something akin."

THE DEATH OF BOB, THE SOUTH AUSTRALIAN RAILWAY DOG.

[*Oct.* 19, 1895.]

Bob, the South Australian railway dog, has ended his eventful career, which is, I think, worthy of notice in the *Spectator.**
Like many other clever dogs, he was of uncertain breed. As a puppy he was attached to a rabbiting party in our North country, and, while still young, was given to a railway guard, with whom he travelled for some time, having been taught to jump into the van, our narrow-gauge lines having no platforms. Bob very soon came to consider himself as one of the railway staff, and although civil to passengers who spoke to him, he never made friends with any but railway employés, whom he seemed instinctively to recognise. The engine-drivers and stokers were his special friends, and for many years he travelled all over the South Australian lines, and occasionally over those connected with them in the other Colonies. His favourite

* See also *Spectator* for September 21.

seat was on the tender, and his whole demeanour showed that he considered him self an important adjunct to the locomotive. He belonged to the department, not to any individual driver, and I have seen him jump off one engine and join another, apparently without any reason, when passing at small roadside stations hundreds of miles from the terminus. His licence was always paid for by the men, and he wore a collar which bore the legend : "Stop me not but let me jog, I am Bob, the drivers' dog." The interest of his career lies in the fact that he attached himself to the locomotives, recognised no individual as master, and no house as home. He seemed to travel from pure enjoyment of movement, and was quite as much at home in the small up-country stations as in the city. He never seemed to be in a hurry, often remaining in the station till the last moment and joining the engine just as it started. He was well fed, and in spite of numerous predictions to the contrary, was not killed by accident on the line, but died in town at a good old age.   ALEX. B. MONCRIEFF.

MORE MISCELLANEOUS STORIES.

A SHEEP-DOG'S MIND.

[*Dec.* 21, 1895.]

WHENEVER I sent the shepherd with sheep to
the local auction the shepherd went in front,
and Turk, a cross between a retriever and
collie, followed leisurely behind. He helped
to put the sheep in the allotted pens, and
then while the shepherd betook himself to a
neighbouring "pub," Turk lay down before
the pens. He always stayed there until the
auctioneer came along and sold the sheep.
Turk watched him carefully as he went
from one pen to the other ; and as soon as
the hammer had fallen on the last pen, he
wended his way to the publichouse, found the
shepherd, and went home with him. Subsequently be became both blind and deaf, and
quite incapable of work. He also took to
coming into the house and lying there ; and
as my children are little, and consider all dogs
their particular playmates, and as Turk's

temper became uncertain, I was obliged to have him shot. I feel sure if I could have explained the matter to him he would have recognised the justice of the decree.

FRED HORNE.

A COLLIE'S INTELLIGENCE.

[*Oct.* 26, 1895.]

A NEIGHBOUR of mine has a young collie which sleeps in the kitchen, where is kept during the night the key of the gate of the yard. The yard-man on his arrival in the morning is accustomed to tap at the kitchen window for the key, which the maid-servant then hands to him through the bars of the gate. One morning lately the maid happened to be out of the kitchen when the man tapped, and the dog (who must have realised the meaning of the taps) took the key in his mouth and carried it to the man at the gate. The dog is very highly bred, but has never been taught to fetch or carry, and is only about a year old.

A READER OF THE "SPECTATOR."

A RELIGIOUS DOG AND PAGAN CAT.

[*Oct.* 26, 1895.]

OF the telling of many stories of cats and dogs there is no end, and much reading of them is a delight to the flesh. Here is a genuine one told to me by a dear and most trustworthy friend—an incumbent in Yorkshire. His dog had certain religious instincts, and when he saw the books brought out for evening prayers, retired to his corner. One evening they were brought out while he was gnawing a bone. Instinctively he dropped it and withdrew. The cat, being a pagan and carnivorous, took possession of the bone. The dog glowered at her, but budged not an inch. Scarcely had the last " Amen " sounded, when he made one spring. The fate of that cat I have not words to describe.

ROBERT GWYNNE.

A PRAYING DOG.

[*Oct.* 26, 1895.]

A FEW weeks ago I sent you a dog story. I
beg now to send you another, related to me
by the Bishop of Wakefield, when he was
rector of Whittington, in the county of Salop.
Dr. How is, I believe, a Shrewsbury man,
and is therefore well acquainted with many a
Salopian family. Well, in Shrewsbury a
certain family had a dog of a religious turn
of mind, who regularly attended the family
prayers. When the bell rang for morning
and evening prayer, the dog invariably
accompanied the household into the room
where prayers were said. Of course, each
member of the family would kneel. down,
leaning upon a chair and with the head bowed
down, supported by the hands and arms.
The dog would copy this example exactly.
He would sit upon his hind-legs, and in that
way copy the kneeling of the family. Then,
in order to copy the arms resting on the chair
and the head in the hands, the dog would
put his forelegs on the chair and his head

down between them. He would remain in this attitude until prayers were over, and then, when the family rose, he would also rise, and perhaps leave the room with some members of the household.

LUDOVICUS.

A DOG'S ADVENTURE.

[*Oct.* 26, 1895.]

MAY I be allowed to add one more to the dog stories which have appeared in the *Spectator?* When my brothers and I were young, we had a white French poodle as our friend and constant companion. He was a strong, muscular dog, standing, I should think, about 18 in. high at the shoulder, and quite the most intelligent dog I have ever known. Among other accomplishments, we had taught him to climb a ladder. He went up very cleverly, and could sometimes turn round and come down; but he could not always depend upon doing this successfully, and occasionally he slipped and came down with a run, but we were always there to catch him, so no harm was done. The dog was inordinately fond of running after stones, and was seldom without one in his mouth. In those days, I am afraid, we were hardly alive to the grinding effect of stones upon the teeth. In the part of Devonshire in which we lived there had been a great deal of

mining for copper, and there were various
workings, old and new, on my father's estate.
In a wood, which stood on the side of a steep
hill, not half-a-mile from the house, a gallery,
or "adit," as it is called locally, had been
driven into the hill-side in the hope of inter-
secting at a lower level a lode which had
shown itself above. To those who passed
down the main path of the wood this adit
showed itself as a cave, quite dark within.
Going that way one day with my brothers
and having the poodle's stone in my hand,
I idly and thoughtlessly threw it into the
mouth of the adit. The dog rushed after it,
and to my surprise and horror, we heard the
stone fall, and immediately afterwards the
dog. This told us that there was a shaft in
the adit, a most unusual thing ; we listened
but could hear no sound, and we had not a
doubt that the dog had been killed ; one
thing surprised us, it was well known to us
that all disused shafts had water at the
bottom, but we could hear by the sound of
the fall that it had not been into water. The
loss of our favourite was a terrible blow, but

we determined, if it were possible, to ascer
tain his fate, and at least to recover his body.
We rushed home, procured the longest ladder
we could find on the emergency, a rope, a
lantern, with a long string attached to it, and
a couple of men. I should think the ladder
was about 22 ft. long, With these we went
to the adit ; on letting down the lantern into
the shaft, there we saw the dog on the ledge
of rock or earth, looking up and apparently
none the worse for his fall. We lowered the
ladder by the rope, one of us intending to go
down and carry him up, but we found the
ladder was not long enough to reach from
the ledge where the dog was standing to the
edge of the shaft ; and this presented a
difficulty which we began to discuss. However, no sooner was the ladder fixed than the
dog began to climb it, and our shouts could
not prevent him. As the ladder did not
quite reach to the edge of the shaft we feared
that when he got to the top he might slip and
have another fall, and this time probably to
the bottom of the shaft, for we could see that
all was dark beyond the ledge on which he

had been standing; owing to some mining freak the shaft had stopped here, but had been sunk again a few feet to the right.   Up came the dog ; the longest of us bent over the edge of the shaft, the others holding on by his heels, he just managed to reach the scruff of the dog's neck, and hauled him up ; and there he was among us safe, and showing every sign of gladness to be with us again.   I can hardly say what form our rejoicings took at the moment, but the dog was a more beloved companion than ever.   He did not show the slightest sign of having been hurt by the fall.

J. F. COLLIER.

THE DOG AND THE MATCHES.

[*Sept.* 21, 1895.]

I HAVE a fox-terrier whose idiosyncrasies excite much interest. Professor Lloyd Morgan, of University College, Bristol, chronicled the same in one of his articles dealing with animal instinct. This dog never sees a match lighted without attempting to put it out, and jumps and snaps at it in a most excited manner. When he was quite young, I dropped something on the floor, and as it was growing dark, lit a candle and stooped down to look for it. The dog jumped at the candle and extinguished it. I thought it was done by accident, and relit it. The animal snapped again at the flame, and again put it out. He has often singed himself subsequently, but has always persevered, when permitted, till he has put out a match lighted and held within jumping reach, or a lighted candle; but as paraffin lamps are used in our house, we have thought it rather dangerous to encourage his proclivity lest it might lead to accident. He also, if a small pair of tongs be taken out of the fireplace and given to him, behaves in a most singular manner, whining over them most plaintively, seizing them in his teeth, and then letting them go again, and whining as if begging them not to hurt him, just as in " Robinson Crusoe," Friday is said to have talked to the gun. We can only account for this by the fact that, when a very young dog, one of the servants threatened to pinch him with the tongs—perhaps she actually did so ; but the reason for his light-extinguishing propensity is totally an enigma to us.

ALGERNON WARREN.

[*Sept*. 28, 1895.]

THE story in the *Spectator* of September 21, reminds me that I once possessed a dog who had precisely the same trick of attacking fire as that mentioned by your correspondent. He was a red Irish terrier that I bought in Kildare when so young that I am sure he had not been taught the trick. He would 'paw" at a lighted match on the ground, or would seize in his teeth a lighted piece of paper and shake it till he had put it out. In the same way he would "worry" at a cigar end thrown on the ground, and never leave it till he was satisfied there was no fire left. I may mention that he once, in Canada, killed single-handed a skunk—an animal which, as a rule, it is said a dog will not face. I wished myself he had not, as for months afterwards his presence was evident long before one saw him, on a wet day particularly.

OLD SOLDIER.

CRIB.

[*Sept.* 28, 1895.]

THE following notes relating to Crib, a white bull-terrier, were dictated by his owner, William Essex, iron warehouseman, who had charge of a horse :—

" Being away for a day, another man was left in charge of the horse. Crib took possession of the stable, and would not let him go in to feed the horse. One of the blacksmiths thought of a plan, went into the next yard and shouted 'Essex!' Crib ran out to see where Essex was, and they shut the door for the man to attend to the horse. Crib frequently went with my fellow-workman, George Harcourt, home to meals. On one occasion he missed him. When he (Harcourt) came back from breakfast, he told the dog he ought to have gone, as he had a lot of small bones for him ; but he must go up to dinner with him. Taking him at dinnertime, he told his wife he had brought Crib to have the bones. She replied, ' You had not been gone ten minutes from breakfast

before he came and had them.' He had
never been known to go there by himself
before. An old man, a Quaker named
Fletcher, lodged with me, and would frequently take Crib a walk. Going across
Merstowe Green the clock commenced striking the quarters for five, which was my tea-
time. At the first stroke of the clock, the
dog stood still, put his head on one side, and
attentively listened till the clock struck five.
With the last stroke, Crib turned round, ran
home, and met me as I went to tea. We
had been at opposite ends of the town. Mr.
Fletcher arriving at home, the first word was
to my wife, ' Mary, what time did Crib
come home?' 'About three minutes past
five.' 'O, beggar him, he knows what
o'clock it is; for as soon as it began striking
he stood still and listened; and as soon as it
had struck the last stroke he ran back home.'
On another occasion I and Thomas Handy
were at work in my cellar. Handy, seated
on the second step, pulled out a packet of
lollipops, asked me to take one, asked Crib
to take one, took one himself, screwed the

paper up, and put it in his pocket. Crib
then left the cellar. In about fifteen minutes
Handy asked me to have another, put his
hand in his pocket, and cried out, 'That
d—d dog 'a got 'em.' Crib had meantime
been up the cellar steps on his left hand side,
picked his pocket unperceived, returned on
his right-hand side, gone into the back
kitchen, opened the paper, which he left there
empty, and quietly enjoyed what he had
quietly stolen. On another occasion we had
young potatoes for dinner. As we could not
mash them with the gravy, Crib would not
eat them, licked all gravy from the potatoes,
hooked them off the plate and placed them
out of sight under the rim. My wife went
into the back kitchen to see if he had eaten
his dinner, and said, ' There's a good dog for
eating the 'taters.' Crib looked up, wagging
his tail, with a ' bow-wow.' As soon as she
stooped to pick up the plate he dropped his
tail, went into the front room, and ran under
the easy-chair out of sight. My wife called
the rest of the family to see the potatoes
in a perfect ring under the edge of the plate.

On Sunday night my wife put my everyday working-jacket in my elbow-chair for Crib to sleep on as usual. He went and looked at the coat, then crossed the room, looked at and smelt my black Sunday coat. My wife asked him, 'Do you want Daddy's Sunday coat?' and he answered with a 'bow-wow.' She took the coat, removed the one that he was in, and before she could place the other, Crib was in the chair. She took the coat, remarking that he could not have the Sunday coat, and replaced the other. Looking very disappointed he jumped down, and remained all night on the cold stones. The undisturbed cushion showed that he never went to his usual bed. Crib always took tea, but would not drink it except from my wife's saucer, which was different from the rest. If it was given in any other he would go and look, but would not touch it till it was put in my wife's saucer. Being a Good Templar I was accustomed to take from home a jug of cold water on 'Lodge' night, Friday. Crib unperceived followed us one night. He was admitted,

properly clothed in the regalia (the broad
ribbon being put round his neck and crossed
over his back), sat very quiet and looked
very pleased for an hour and a quarter.
From that time we could never keep him
from ' Lodge.' Afterwards when the jug was
placed on the table before starting from
home, if the door was open, he would immediately start and go to the lodge room in
the next street. Crib's master was caretaker
of the Friends' Meeting House, the door of
his house opening into the passage up which
the Friends had to pass. Crib would lie still
and take not the slightest notice whilst the
Friends belonging to Evesham went up the
passage. Should a stranger be with them,
Crib would bark the moment an unaccustomed step was heard. At one time there
was something wrong with Crib internally.
When the pain came on, he would set up his
back, go round and round and cry out most
piteously. I was recommended to give him
laudanum. When he found the pain coming
on, he would stand and look up at the bottle
on the shelf, then look at my wife or

daughter, then at the bottle, jump up in the big chair and lie quiet for a dose of laudanum. This he did twenty times. Poor Crib went mad, and had to be destroyed in his eleventh year, September, 1874."

D. DAVIS.

A CLEVER HUNTER.

[*Sept.* 28, 1895.]

As your readers seem interested in stories of
canine sagacity and cleverness, I gladly send
you a short account of a small spaniel's
singular action and acuteness of thought. A
few days ago I was taking a walk before
breakfast in some fields near my house,
accompanied by my little dog. I did not
pay much attention to her doings, but noticed
she was running about as if in search of
game. However, on my way home I found
the dog was unwilling to follow me. She all
the time wished to turn back. She would
follow a few yards behind if I went on ; but
if I looked round she would immediately
pause, and then make her way back towards
the fields. This happened several times
At last I concluded that the spaniel had
some object in view in wishing to retrace her
steps, and so I returned with her, she
leading the way and I following. She went
straight to a rabbit, and bolted it. We had

a good chase, and at last succeeded in catching the rabbit.

Now, the dog had evidently discovered the rabbit on its form when ranging about the fields, but thought it unsafe to start it in my absence, for I had left the fields and was now on the high-road. She clearly wanted my help and encouragement in the chase. I would observe that we have here an instance of great caution on the part of the dog. Her natural impulse would be to start the rabbit at once and pursue it. This impulse the dog checked. Moreover, I would point out that my little bitch seemed to exercise her reasoning powers, and that in a marked way. She, as it were, said to herself:—" I will not bolt the rabbit in the absence of my master. I will run after him and bring him back, and then, encouraged and helped by him, I shall start the rabbit, and, if possible, catch it." I consider that my little dog showed that it possessed the faculty of reasoning in checking its natural impulse, which would lead it to spring at the rabbit at once, and also in fetching me back to be a witness and a helper in the chase that ensued. All her actions manifested caution, sagacity, and the possession as well as the exercise of the faculty of reason.

LUDOVICUS.

THE HOMING INSTINCT IN DOGS.

[*Aug.* 10, 1895.]

AMONG your numerous dog stories perhaps the following may find a place. I have a Skye terrier puppy, only nine months old. On Thursday afternoon my son and a friend took him from here outside an omnibus to Coleridge's village, Nether Stowey, nine miles nearly due west. They then walked to another village, Stoke Conrey, three miles to the north. Leaving him outside the church for a few minutes, he had disappeared when they left it, and the only trace of him that could be found was the report of some men who had seen him running over a hill still further to the north. On Friday night, at 12.30, he reappeared at home. He must have either retraced his steps to Nether Stowey, and then come home by the road the omnibus went by, two sides of a triangle, twelve miles, or else come home by the main road from Stoke Conrey, a most complicated and winding road, nine miles, which he had never seen before.   Either feat seems rather startling from such a canine baby, and makes his name, " Teufel," rather appropriate.

E. T. PAGE.

INDEX.

www.ingramcontent.com/pod-product-compliance
Lightning Source LLC
Chambersburg PA
CBHW021500210326
41599CB00012B/1071